Inquiry and Understanding

Also by Jennifer Trusted:

The Logic of Scientific Inference (Macmillan, 1979)
An Introduction to the Philosophy of Knowledge (Macmillan, 1981)
Free Will and Responsibility (Oxford University Press, 1984)
Moral Principles and Social Values (Routledge and Kegan Paul, 1987)

Inquiry and Understanding
An Introduction to Explanation in The Physical and Human Sciences

Jennifer Trusted

**MACMILLAN
EDUCATION**

First published 1987

Published by
MACMILLAN EDUCATION LTD
Houndmills, Basingstoke, Hampshire RG21 2XS
and London
Companies and representatives
throughout the world

Typeset by Wessex Typesetters
(Division of The Eastern Press Ltd)
Frome, Somerset

Printed in Hong Kong

British Library Cataloguing in Publication Data
Trusted, Jennifer
Explanation.
1. Inquiry and understanding.
An introduction to explanation
in the physical and human sciences.
I. Title
121'.4   BD161
ISBN 0–333–43958–9
ISBN 0–333–43959–7 Pbk

Contents

Acknowledgements

I should like to thank Professor D. J. O'Connor for his advice and encouragement in the writing of this book; I should also like to thank Professor J. Harrison for being kind enough to read the manuscript and for some invaluable suggestions for clearer exposition. My thanks also go to Mr Martin Davies for his helpful guidance in relation to presentation and style. Any mistakes in the text are, of course, my own. The author and publishers wish to thank the *Daily Mirror* for permission to reproduce their horoscope feature of 25 July 1986.

JENNIFER TRUSTED
Exeter, 1986

Preface

This account of empirical explanation is intended to serve as an introduction to philosophy, especially for those interested in the philosophy of the physical and human sciences. I hope that it will also be useful to scientists themselves. No prior knowledge is assumed and the text is written for the general reader as well as for students.

A realist view of empirical theories is adopted, that is true empirical theories are held to describe an objective reality, though any claim to knowledge and any explanation are, and must be, subject to the limitations of the human condition. Our knowledge of the world is taken to be objective knowledge in the sense that evidence provided by our observations, and the interpretations of that evidence provided by our theories, is accepted by those who are capable of assessing such evidence and such interpretations. We know that we may be mistaken, but we can admit this without resort to scepticism; at the last we must be content to rely on our own judgement, including what we judge to be inter-subjective agreement, for we have nothing else to go by. This is not to undermine the notion of objective truth but to point out that we can never be *certain* that we have attained truth. Hence it is not denied that probably all our currently accepted empirical theories are likely to be modified and that some will be rejected as being false, or highly misleading. Nevertheless it is argued that we can hope to make progress towards the truth as we human beings can understand it.

In a simple introductory account it seems best to assume (by implication) a correspondence theory of truth and to treat empirical propositions and statements as being either true or false. Arguments for anti-realism (as presented by Dummett) are ignored, as is the problem of the connection of meaning with truth conditions and the case for intuitionist as opposed to two-valued logic. A detailed account of the interdependence of our notions of truth, meaning and use, and their relation to the metaphysical assumptions we make about the world is beyond the scope of this book.

The first chapter is a general introduction and presents a common-sense account of the nature of empirical explanations, distinguishing them from logical and metaphysical explanations. It is suggested that the concept of explanation is logically connected to the concepts of knowledge and understanding and that explanations provide descriptions, suggest causal links and place the *explicandum* in a broader scheme of events. There is also preliminary discussion of the role of teleological explanations and of the operation of *Verstehen*.

Chapters 2 to 4 are intended to provide a philosophical foundation for the arguments developed in later chapters and there is constant reference back to these early chapters. In them it is argued that facts, concepts and theories are inter-related and that the meaning and significance of words and propositions is dependent on concepts and theories, so that language changes as concepts and theories change. The treatment is simple, for any deeper approach would require background knowledge which neither the student nor the general reader could be expected to possess. It is hoped that though the simple account offered does ignore many important philosophical problems, it will not mislead but will provide a sound basis for further study and will not confuse the beginner. Some of the simplified ideas will need to be modified later just as in the teaching of physical science the student learns of simple laws and theories that also have to be modified later.

From the first chapter there is implicit acceptance of the importance of laws for explanations; laws are taken to be established regularities and the well-known difficulties associated with the status and reliability of laws are treated later. At the start it is assumed that the reader is like Hume out of his study, and common-sense reliance on inductive inference is assumed. It is in Chapter 6 that the problem of induction is introduced and briefly discussed, along with the difficulties inherent in distinguishing laws from contingent true generalisations; in Chapter 8 the notion of quasi-laws is developed. It is not pretended that these discussions are deep; they are intended to alert the reader to interests that will arise if she perseveres with philosophy.

It is argued that the D-N model of explanation is a model for the causal aspect of explanation as offered by common sense and by science. Physical science (rather than 'natural science') is contrasted with human science (the latter term embraces 'social science') to emphasise that explanation in the former appeals solely to physical events whereas explanation in the latter can involve appeal to intentions and beliefs (see also note 10, Chapter 1). It is admitted that there is more difficulty in applying the D-N model in human science explanations but it is held that those who have criticised the model have not given sufficient weight to the necessity to appeal to some established regularity; even a weak law is better than no law at all. It is argued that teleological explanations, in so far as they are treated as causal explanations can also be accommodated to the D-N model.

However, it is also contended that the D-N model is inadequate as a model for full explanation in both physical and human science because it ignores the conceptual aspect of explanation and allows no place for changes in theories, concepts and language. It is pointed out that the inadequacy is revealed more clearly in the human sciences since in those fields we are liable to have complex and fluid concepts that are deeply influenced by values – explanations and theories about ourselves are inescapably value-laden and even the operation of *Verstehen*, that encourages us to understand the viewpoint (and values) of others, cannot entirely eliminate subjective assessment and subjective explanation.

The final chapter summarises the main arguments of the text and reviews the possibility of an ultimate explanation, in the Baconian sense. It is suggested that not only is it highly unlikely that there is an ultimate explanation, or an ultimate set of explanations, but that were it (or they) known there would no longer be a place for the exercise of our human aspiration to discover more about ourselves and our world.

Empirical Explanations

In this book we are concerned with the nature of explanations about the world around us, that is with empirical explanations. We can study the properties and interactions of many kinds of phenomena: earth, air, fire and water; solid material objects, magnetic and electrical fields of force, shadows, mirages, mirror images, atoms, and electrons, stars, quasars, and also plants and animals, including ourselves.

The behaviour of all these phenomena is susceptible of *empirical* explanation because we use our senses, and must use our senses, to learn about that behaviour. Our observations may involve one or more of our senses: looking at a star; hearing the bell toll; touching or 'feeling for' an electric switch; smelling a gas leak; seeing, touching, smelling and tasting a lemon. Sometimes observation is indirect; necessarily so when we detect a magnetic field with a compass, but also when we 'see' an aeroplane by observing its vapour trail, hear a voice on the telephone, prod (and so feel indirectly) a patch of soggy ground. There is no absolute distinction between direct and indirect perception: do we say that we see something indirectly if we use spectacles, or a microscope or a telescope? Do we hear indirectly if the speaker reaches us with a microphone or we 'listen' with a stethoscope? Yet the distinction between direct and indirect perception can be useful and so is worth making – we distinguish black from white, and find the distinction useful, even though we can pass from black to white through shades of grey.

Empirical explanations are to be contrasted with explanations in logic or in mathematics; for example an explanation showing that there is no greatest prime number is not an empirical explanation because it is not dependent on an appeal to observation but on an appeal to reason. Diagrams and models may be used to help understanding of a mathematical or logical problem and to arrive at a solution and/or an explanation but there is no necessity to invoke evidence based on observation. Empirical explanations are also to be distinguished from metaphysical explanations which involve assessing the compatibility of our basic assumptions and beliefs, those that are presupposed by all our empirical explanations. These beliefs are generally left unexpressed but they are held very firmly and we find it almost impossible seriously to question them. They provide the framework that gives us our empirical world and, in this text, the basic metaphysical assumptions and beliefs that support the view that there is an external world independent of our thoughts and perception, that there are physical objects located in a three-dimensional space, that physical events may be explained by causal laws and that there are other people with thoughts and feelings similar to our own will not be questioned.

Empirical explanations are sought, primarily, because we want to understand the world around us. We want to know how things happen and why they happen as they do; we want to understand how and why the unexpected, the unfamiliar and the surprising occur and, rather less urgently for most of us, how and why the familiar routine things happen. The search for explanation is a search for *knowledge* and *understanding*. Explanations are held to be satisfactory if they help us to answer the questions prompted by natural curiosity and there is a logical connection between the concepts of knowledge and discovery and the concept of explanation. Likewise there is a logical connection between the concept of understanding and the concept of explanation for if an explanation is satisfactory it must make the world to some extent more understandable.

As von Wright says,[1] practically every explanation can be said to further understanding, but he considers that the word 'understanding' has psychological overtones that the word 'explanation' lacks. Understanding human behaviour involves the inquirer in feeling as the agents feel, that is she must seek empathy with those she studies. It is also connected with *intentionality*, and this involves knowing the aims and purposes of what the agents *do*. The former underlies the operation of *Verstehen* (see below and Chapter 13), the latter is related to the meaning and significance of conventions, signs and symbols (see Chapter 7, p. 66). In this text both these are treated as part of the 'Wherefore' aspect of explanation (see Chapter 7).

Some, but not all explanations imply excuse and/or justification; generally these are explanations of human behaviour.[2] For example, if we are asked why a person is angry, the explanation may well serve to excuse and even to justify her anger. In addition, explanations are sometimes needed to communicate with others: the scientist may need to explain the significance of certain theories to the non-scientist, members of one society may wish to explain the role of their customs to members of another society. But these two aspects of explanation, justification and communication, are peripheral, they are not central to the concept of explanation and are not logically connected to it.

In due course an explanation may be shown to be false; for example, in the nineteenth century the explanation of the spread of diseases such as cholera was that gases rising from decaying organic refuse, rotting meat, vegetation and excrement were poisonous and carried diseases. Today we think that this is incorrect so the knowledge thought to have been acquired in the nineteenth century was illusory. But though that explanation is no longer accepted it had served as an explanation increasing knowledge of diseases and epidemics and, because it encouraged cleanliness and the introduction of public health regulations, it had great practical use. It had been useful to relate the spread of disease to the presence of dirt even though the relationship had not been properly understood.[3]

Explanations are of three kinds:

(1) The expected, the familiar or routine is related to something new.
(2) The unexpected, the unfamiliar or surprising is related to something already known.
(3) The unexpected, the unfamiliar or surprising is related to something new.

Let us consider some examples; first from the physical sciences:

(1) The passage of the sun across the sky is explained by the daily rotation of the earth as it moves round the sun.
(2) An eclipse of the sun is explained by the shadow of the moon falling on the earth.

(3) Radioactivity is explained by the disintegration of what had been thought to be stable and indivisible atoms.

Take the first example: for a very long time it was thought that the apparent movement of the sun across the sky was due to the sun actually moving round the earth. This seemed obvious to common sense, though some people did appreciate that if the earth rotated it would look to us as if the sun were moving. But the earth appeared to be massively immobile and the notion that it was whirling round like a top seemed far-fetched and ridiculous.[4] Aristotle constructed a complete cosmology with the stationary earth at the centre of the universe; this was elaborated by Ptolemy about AD 120 and became part of accepted cosmology and natural philosophy until the sixteenth century. In 1543 Copernicus published an alternative explanation which we now take to be as obvious to common sense.[5] Copernicus has the credit for the suggestion that the earth is rotating because he supported his conjecture with evidence based on observation of the positions of the heavenly bodies and he showed how, on the basis of his new cosmology, these positions could be predicted much more accurately than before. There were many objections to the Copernican cosmos, objections which were valid in relation to sixteenth-century science (the current knowledge of the physical world). The objections could not be met until the next century when Galileo was able to propose a different account of physical movement. Nevertheless the Copernican scheme invited serious consideration from the start.

Most of us do not seek explanations of the familiar and expected; we are irritated by the persistent questions of small children asking for explanations of what we take for granted. Children look at our world with a vision less obscured than ours by the dogmas of common sense and their questions invite us to look critically at what we take as familiar and obvious and to reassess that comfortable cosiness which pushes the inexplicable outside the boundaries of our everyday world. Many of the greatest discoveries, like that of Copernicus, have come about because a genius transcended common-sense boundaries and questioned that which contemporaries thought posed no problems. It is for this reason that a new account of what had seemed to be already adequately explained will call for a reassessment of familiar facts and can be more revolutionary than the discovery of entirely new phenomena. As we shall see, acceptance of the Copernican explanation of the apparent movement of the sun entailed accepting new laws of motion, new theories of force and indeed, a new physics, as well as a new cosmology.[6]

The examples given above are, as was stated, examples of explanations of physical events; we may compare them with explanations of events that include human actions:

(1) A small boy's naughtiness is explained by appeal to an Oedipus complex.
(2) An outbreak of rioting is explained by appeal to frustration caused by poverty, unemployment and racial discrimination.
(3) The fact that there has been no major war since 1945 is explained by appeal to the fact that the great powers possess nuclear weapons.

These examples are parallel to those of the former group but apart from their content, we see two major differences; firstly the explanations offered do not command general acceptance and secondly they depend on appeal not only to physical events, but also to human desires and emotions. Lack of agreement is partly due to the complexity of situations and the lack of knowledge of relevant facts but it is also due to our inability to make accurate predictions of how human beings will behave and our inability to know what their desires and intentions are. We have to infer goals by observing behaviour (which of course includes speech) but we know that we can be mistaken; our 'laws' relating overt behaviour to desires are not as reliable as physical laws.[7] However, I suggest that the principal reason for disagreement, and for passionate disagreement, is that explanations of human behaviour almost always involve implicit (and sometimes explicit) value judgements. If some people think that certain kinds of behaviour are immoral they will object to explanations that do not carry condemnation of that behaviour; if they think that certain individuals are morally weak they will reject explanations that do not depend on appeal to that weakness; if they think that an explanation ignores (and a fortiori undermines) their values and perhaps, less nobly, their interests, they will find it hard to accept and will seek alternative explanations. Hence the diverse explanations offered for a large number of social and personal events: the outbreak of wars, inflation, road accidents, alcoholism and so on. Nevertheless we do not abandon the search for explanations in human affairs and the hope that there can be some measure of agreement. We do, in general, accept certain psychological and social 'laws' as being sufficiently reliable to use in explanations; for example 'Small children are disturbed if they cannot see their mothers in hospital' and 'If goods are scarce prices rise'. Even proverbs and the humble maxims of common sense may serve as laws supporting explanations: 'The burnt child dreads the fire' and 'A stitch in time saves nine.' Psychological and sociological laws undoubtedly show more exceptions than do the laws of physical science but at least to some extent this is a difference of degree rather than of kind. For there *are* exceptions to scientific laws; for example a simple accepted scientific law, 'Matter expands on heating and contracts on cooling', has some exceptions, the most well known being the behaviour of water near its freezing point. This exception, along with others can be subsumed under a more comprehensive law that gives us greater knowledge. Likewise we hope to be able to subsume the unusual behaviour of people and societies under more comprehensive laws and so to arrive at more reliable laws and thereby gain more knowledge of human beings both as individuals and as members of social groups.

Just as general physical laws give us our concept of inanimate materials and non-sentient objects so psychological and sociological laws give us our concepts of human beings – through the laws we establish we learn more. There are philosophers, psychologists and sociologists who think that we should be able to dispense with laws appealing to conscious purposes and that ultimately all human actions are explicable solely in terms of physical laws. Such laws would describe brain and other bodily happenings and would relate overt behaviour to neurological and chemical changes; the behaviour of people in social groups would be explained in terms of evolutionary laws, themselves ultimately reducible to physiology and then to chemistry and physics. These philosophers think that our inner consciousness, our thoughts, feelings and desires, have in themselves no direct effect on behaviour and are but epiphenomena. If this were so it would follow that all psychological explanation must ultimately be superseded by physical explanations for these would be fundamental. They would also be more reliable since physical laws can be more securely established than can laws appealing to conscious thoughts.

The view taken here is that the reduction of explanation in terms of conscious thoughts to physical explanations is very unlikely to be possible even in principle; clearly it is not yet possible in practice.[8] For not only is it not possible in practice to explain human actions in terms of physical events and physical laws alone, it is at least arguable that human actions are *not* entirely caused by purely physical events. But even if they were so caused, we do not necessarily want to know the preceding bodily events when we seek explanations of human behaviour; our curiosity about the behaviour of individuals and of societies is almost always a curiosity about their goals, their motives and their desires.

So much is this a feature of how we view the world and, in particular, living things, that we tend to give similar explanations of the behaviour of non-human animals, and even plants. It may be conceded that the reference is largely metaphorical (see also Chapter 10) but there is no doubt that we find accounts in terms of goals and motives satisfying. We speak of leaves turning *in order to* get the sunlight, of bees dancing *in order to* mark territory, of birds migrating *in order to* reach a region of equable temperature, and so on. In the case of higher mammals this kind of explanation seems the most natural. We know that these animals can learn and that the learned behaviour can be passed on, through the offspring imitating the parents, to subsequent generations. Trigg cites the well-known case of the Japanese monkey who washed sandy potatoes with sea water. Other members of the tribe imitated her and now most of them do this.[9] The animals liked the taste and it is not unreasonable to assume that they wash the potatoes with the conscious intention of making them more tasty. Evolution provides some justification for us to believe that at least some animals are motivated as we are. Of course it is possible that they are not but it does not then follow that they are automata. In so far as they have motives and intentions that we cannot infer, their behaviour cannot be explained; we *have to* explain in human terms, nothing else is possible for us.

When we seek explanations of human actions we are on firmer ground for though we differ from each other the differences are superficial as compared with the differences between humans and other animals. Therefore we are in a far better position and are entitled to have some confidence in our inferences. We may be helped by the operation of *Verstehen*, the process of understanding a situation from another's point of view, for if we appreciate how other people assess their own circumstances we shall be in a better position to explain their behaviour. In carrying out the operation of *Verstehen* we do not primarily seek to relate that which is to be explained (the *explicandum*) to some accepted regularity or law; we may do this incidentally but, as is to be shown in Chapter 13, the purpose of the operation of *Verstehen* is to gain understanding in von Wright's sense (see above, p. 2), that is to see the situation as the actors see it. We need to know not only what they think is right and wrong, beautiful and ugly, polite and impolite but also to appreciate those evaluations. To have this understanding does not entail that in their circumstances we would decide to act in the same manner; we may not accept the alien values and/or the assessment and evaluation. We may think that the *explicandum* neither justifies nor excuses the behaviour for it is not necessarily the case that *tout comprendre est tout pardonner*. But we must undertake the operation of *Verstehen* to be in a position to claim a right to reject alien assessments.

Here it will be argued that satisfactory explanations in the physical[10] and in the human science will not only relate the *explicandum* to some regularity or law but will also give understanding and that the operation of *Verstehen* is essential to understanding and essential for a satisfying explanation of human behaviour.

Facts and Concepts

To start our analysis of empirical explanations we have contrasted explanations in the physical sciences, which invoke appeals solely to physical laws as part of the *explicans*, and explanations in the human sciences which, though they may well invoke appeals to physical laws, also involve appeals to psychological and sociological laws expressed in terms of human feelings and intentions. As we saw in Chapter 1, the examples of explanations given there are open to dispute partly because they rest (implicitly or explicitly) on appeal to values and partly because the laws invoked apply to complex situations where evidence is generally inadequate. We saw that there were also peculiar difficulties involved in inferring intentions and goals. There is yet another difficulty, namely that the laws may be imprecise and misleading because the words used refer to concepts that are not clearly defined. For example we have the law 'Poverty causes despair', but what is meant by 'poverty'? We have the law 'Racial discrimination causes frustration' but what is *racial* as opposed to other sorts of discrimination? We have the maxim 'Major powers are open to blackmail by terrorists' but what makes a major power and how do we define 'terrorist'? Different people have different concepts and may use the same word in different senses; so they fail to understand each other and, in particular, they fail to understand each other's explanations. The following letter from *The Times* makes the point explicitly; the writer is referring to a dispute about Health Service salaries:

Sir,

Mr Frank Chapple, no liar, is deluding himself when he says that everyone supports the health workers except Mr Fowler. It is just that few of Mr Fowler's supporters can be heard from inside a trade union headquarters. The case that we ought not to devalue skill and so must not pay more to the lower-paid workers only, and that the price of a greater rise all round is more than the rest of us can afford, is almost inaudible from there.

We must alleviate hardship, but hardship is children without shoes, Lowry people with too little to eat, damp walls and leaky ceilings. It is not involved in cutting out smoking, drinking, betting and holidaying abroad.

Almost the only avoidable hardship after 35 years of reasonably good government affects two classes of people, those overlooked as individuals by imperfections in the administration and those who are robbed of what they are entitled to, such as the sick, by people who are trying to take more of the inessentials of life than they are able to obtain without menaces.

The same point is made by Bernard Levin; in the article below he is, of course, implying that the criteria of poverty adopted by certain people are inappropriate in that they are based on a false concept of poverty. We are not concerned with the political implications of the article, it is the philosophical point about the concept which concerns us:

Poverty they call it . . . that's so rich

They order this matter better in Copeland; or if not better, then without doubt differently. Copeland is in Cumbria, and not so long ago, it seems, the local council sent to ask those of its tenants who were behind with the rent why this was so; the problem was a pressing one, apparently, because nearly half of all those dwelling in municipal property in the area were in arrears, and the resultant hole in the municipal books has had to be filled up, come ratepaying time, by the other half.

Only two of the answers given on the doorstep to the man in the bowler hat were published; in those two replies, however, there rests much matter for wonder. One family (the breadwinner was earning some £7,500 a year) said that they could not afford to pay the rent because they were already paying £25 a week for the hire of five television sets and three video recorders. Another family in the area had got behind with the rent because of the cost of a summer holiday they had taken in Algeria; when the collector ventured the opinion that that must have set them back a bob or two, they explained that the Algerian trip had been necessary because although they had already had one holiday that year, in Malta, it had rained while they were there. The council (Labour-controlled, incidentally) thereupon took steps to regain possession of the rent-owing families' houses.

. . . .

The trouble began, I think, in the use of the word 'poverty', and the reason it caused trouble lay in the fact that it cannot be defined except in relative terms. An unemployed and partly disabled elderly woman living in one room of a condemned tenement in the Gorbals would, I think, be held to be poor by any reader of these words. But to a family living on the pavement in Calcutta the Gorbals woman is a Maharanee dwelling in fabulous luxury. So much is obvious (though you would be surprised at how widely it is not understood); what is less obvious is that the usual answer to the point implied in the comparison – that the Gorbals woman does not live in India but in a country where most people live in decent houses or flats – won't do either. For what, under the new dispensation, does the Gorbals woman need to be no longer poor?

Certainly she needs the leaky roof mended; she needs more and better food; she needs heat, clothes, washing facilities. But that is what she needs to avoid breakdown, starvation or hypothermia; what does she need to be no longer thought of as poor? It may be difficult to believe, but there is no possible answer to that question.

In 1982 the proportion of households in Britain with a television set was 97 per cent; were the other three per cent poor? It seems they must have been, for to lack what almost everybody else has is the accepted definition of poverty. Then a television set is a necessity. But wait: the 97 per cent of households with a television set were divided into 77 per cent with a colour set and 20 per cent with a black and white. Not to have what three-quarters of the population do have must be to live in poverty; then a colour set is a necessity. Is that not an odd conclusion?[2]

Not every reader of this book will agree with the views expressed by the writer of the letter or by Mr Levin but what is to be stressed here is that any discussion about poverty, whether it ought to be relieved, and at what cost to others is useless, and may be dangerous, if the disputants have different concepts of poverty and have failed to appreciate that this is the case. Similarly any explanation of poverty depends, at least in part, on the concept of poverty assumed by those who seek for or offer an explanation.

Concepts are general ideas, general notions of classes of objects, qualities or events. Some philosophers hold that we show that we have grasped a concept when we demonstrate that we can recognise a given object, quality or event for what it is (a member of its class); others hold that a concept is a capacity for certain exercises of the mind. But on either view we cannot discuss, let alone explain, an event or a state of affairs without the relevant concepts. So, only if we have a concept of poverty can we write or talk or discuss the problems of poor individuals, poor homes or poor districts in a sensible way.

Now it is very likely that any two people will have a different concept of poverty, so they are likely to disagree as to whether a particular individual, home or district is poor. The question 'Is that person poor?' will be answered 'Yes' by one and 'No' by the other. So there will be disagreement as to whether the statement 'That person is poor' is or is not true and does or does not assert a fact. Because different significances of words reflect different concepts there are different opinions about facts as well as about words. If concepts are fluid, facts are 'soft' and any discussion about facts becomes futile if conceptual differences are ignored.

It is not difficult to appreciate the potential softness of facts pertaining to the human sciences because it is clear that what is observed has to be interpreted and evaluated. But the need for interpretation is not confined to these kinds of facts, that is facts depending on complex concepts such as poverty, racial discrimination, major power or terrorist. Much more simple facts, that *seem* to be directly accessible through sense perception are not as 'hard' as is customarily supposed; they are not directly 'delivered' by the senses, rather they depend on our interpretations of what we receive through our sense of sight, touch, hearing, taste and smell. Our sense experiences, considered as raw data, can tell us only about ourselves, not about the world outside; we have to use those experiences by interpreting them.[3] This is clear enough if we consider sensations of taste and smell: for example we are aware that we interpret when we say that a sharp taste is of lemon juice and when we say that a familiar aroma drifting from the kitchen is breakfast coffee. Although the need for interpretation is less obvious, it is still not very difficult to appreciate that we have to interpret sounds and tactile sensations: the ringing is inferred to be a telephone bell, the square button is held to be an electric switch. Such sounds and 'feels' are given significance on the basis of interpretations of sensation. But it is quite difficult to appreciate that visual experiences must be interpreted for we do not seem to need to interpret the 'look' of, say, a table – we see the table, it *appears to be* directly 'given'. This is true of the majority of our visual experiences; it is only if conditions make vision difficult or if what is seen is unfamiliar that we are conscious of interpreting the visual sensation. Yet even our most familiar and clear visual sensations must be interpreted before we see objects and events.

We have gustatory, olfactory, auditory, tactile and visual sensations[4] but *what* we taste, smell, hear, touch and see depends on our interpretation of the relevant sensations. We each make our own interpretation, dependent on our background knowledge and past experiences, and each person's description, the facts expressed, will be characteristic and potentially unique.

It is because we have what is probably an instinctive tendency to believe in an external world of physical objects and people that we instinctively assume that the external world is *responsible for*, that is, is the cause of most of our sensations. At a very early age we form concepts as to the nature of the world, that is concepts of the objects and people in it. Our knowledge of the world is embodied in our concepts; without concepts we could have no knowledge and, *a fortiori*, no understanding and no desire to learn and to understand more. Learning to interpret involves grasping concepts: for example, to interpret a visual sensation as being the sight of a table, we have to possess a concept of tables. The English word 'table' is not important, the concept is. For though we needs must discuss concepts in some language we do not need any language in order to possess concepts, rather we need to have grasped at least some concepts in order to be capable of speech.[5]

Because our concepts do not seem to be innate, that is since there is no evidence that we are born with knowledge of what the world is like, it is difficult to understand how we start classifying and so interpreting our sensations. It has been suggested[6] that we have a natural tendency to mark 'same again' and so to group into classes. Since there is likely to be mental similarity analogous to the physical similarity we observe between people the classifying schemes of different individuals will, though potentially unique, have a great deal in common. This will be reinforced by a common language.[7] As Quine says:

> Different persons growing up in the same language are like different bushes trimmed and trained to take the shape of identical elephants. The anatomical details of twigs and branches will fulfil the elephantine form differently from bush to bush, but the overall outward results are alike.[8]

It is likely that *all* human beings (whatever their language) pick the same characteristics to make their *basic* classifying scheme. As mentioned above it seems likely that we humans (and perhaps some other animals) have an innate tendency to order our sense experiences to give ourselves a world of three-dimensional physical objects in a three-dimensional space. The many different languages of different human societies all depend on these concepts and are evidence that our fundamental classifications are the same. It is because we all do have (or it seems that we have) the same basic concepts that there is a possibility of translating and that there can be communication between speakers of different languages.[9] It *may* be that animals have a totally different classifying scheme; this is the reason that Wittgenstein wrote that if a lion could speak we would not understand him. There can, of course, be very great difficulty in understanding another *human* language and, as has been shown, the same word may express different concepts for speakers of what purports to be the same language. Failure to 'speak the same language' be it due to different words and grammar or to different concepts entails failure to communiate and the latter type of failure can be more dangerous because it may not be appreciated *as* failure. Nevertheless, usually it is possible to make some kind of translation, and therefore to communicate with and understand human beings from different societies and with different cultural backgrounds. Awareness of the difficulties, and of the possibility of failure, makes it more likely that we shall be able to achieve understanding through translation.

What we have to recognise is that all human concepts (and

therefore all human languages) are not so alike that language differences are merely differences between sounds (or marks) – that is between spoken (or written) words for the same 'thing' or 'idea', between sentence construction (e.g. adjectives before or after nouns, or different verb placing) and between superficial grammar (tense endings and genders). Though all languages show that the *fundamental* human classifications (based on concepts of physical objects and causal interactions) are universal, they also reveal deep differences in concepts which reflect different interpretations of the sensations caused by the world outside, that is different interpretations of perceptual experiences. As O'Connor says, referring to the work of the American linguist Sapir, the evidence suggests 'a very close causal relationship between a man's native language and his conceptual interpretation of his environment'.[10] He adds that enquiries such as those of Sapir, and his pupil, Whorf, support the view that 'facts in so far as they are knowable are never concept-free'.[11]

Whorf himself says:

We are inclined to think of language simply as a technique of expression, and not to realize that language first of all is a classification and arrangement of the stream of sensory experience which results in a certain world-order.[12]

Whorf studied the languages of Mexican and American Indians. He showed that the Hopi Indians have no words that refer directly to what we call 'time', or to motion (as moving through space) – there is not even *implicit* reference to time, conceived as a flow of time. Thus their metaphysical framework for interpreting events is different from ours; for example they do not have plural forms for units of time such as 'day'; that is they do not count *days* as 1, 2, 3, etc., instead they refer to day sequences, and to the members of the sequence as 1st day, 2nd day, 3rd day and so on. It has been surmised that rather than thinking of several days (and therefore a given time interval, such as a week) as an aggregate of several smaller time units, following each other, the Hopi think of time as recurring in cycles. It has been discovered that their ceremonies and rituals involve preparation for the repetition of what we would call 'future cycles'. This is, of course, very different from our view of the future; we speak of making a *fresh start*, of New Year resolutions and so on. For the Hopi the future is, in their sense, already there. Their scheme may seem weird and almost mystical to us but it is as capable of accounting for experience as is our own.[13]

There are many other examples of interesting differences between English (and other European languages) and the languages of more exotic societies. The differences of language may be a consequence of different modes of life which make different aspects of the natural world important and so demand different kinds of descriptions. Some American Indians have different verb forms which differentiate hear-say from eye-witness reports, something that might be salutary discipline for our media reporters and certainly of service to those whom they inform. Eskimos have several different words embraced by our one word 'snow', which allows them to talk of different kinds of snow without recourse to qualifying adjectives. Other societies have a language with several different words for water, thereby distinguishing still from running water, and so on. Each language reflects the concepts of its users, and these concepts are the basis of the facts which each society wants to, or feels it needs to be able to, express.

It may be thought that though differences in cultures must arise from different schemes of classification within classes, the very basic concepts, for example concepts arising from visual sensations, do not depend on any intellectual interpretation of those sensations. It is obvious that very small children have concepts of familiar objects such as chairs and tables and teddy bears, and of familiar materials such as milk and sugar. These concepts are partly based on other senses, but recognition is primarily, or to a very large extent, through visual perception. Simple visual experiences (as was mentioned earlier in this chapter) do *seem* to be 'given'; we are not consciously aware of any process of classification or of inference – certainly we are not conscious of any intellectual assessment.

Our dependence on concepts to make even our visual sensations significant is shown by the work of von Senden who, in the 1930s, investigated the process of 'learning to see' which those who had been blind, or nearly blind, from birth (or from a very early age) had to go through after operations such as the removal of a clouded lens (cataract). After the operation these people could have their retinas stimulated by light, and impulses could pass to their brains along their optic nerves. In general they were not mentally retarded, and some of them were exceptionally intelligent.

On the basis of his investigations, and his knowledge of the findings

of others, von Senden concluded that blind people (that is those who
have *never* seen[14]), had no concepts of space and shape as we, sighted
people, conceive them. Our concepts of space appear to arise
primarily from *visual* sensations and, contrary to what had been
surmised earlier, our sense of touch does not provide the experience
necessary for the normal (i.e. sighted person's) concepts. Von Senden
comments on various individuals, for example:

> prior to operation he possessed no idea either of horizontal and
> vertical or of round and angular, but only the tactual relation
> between upright and horizontal, and the distinction between round
> and angular tactual sequences; no absolute spatial concepts, but
> only relational concepts, ordered sequences and schemata.[15]

He cites the case of a patient 'who claimed to have an exact notion of a
horse, and then took a large ten-litre bottle for a horse at a distance of
one foot'.[16] Psychologists and observers, says von Senden, *seem* to
speak the same language but the concepts, and therefore the
significance of the words they use are very different: 'In these visual
tests, indeed, it becomes especially manifest that although both
parties – the observer and the subject – speak the same language, they
attach different concepts to a great many words,'[17] and

> all the spatial impressions given by sight present something utterly
> new and for long unintelligible to the patient, so that it becomes
> obvious that the terms 'space', 'shape', etc., have an entirely
> separate meaning for the blind, which cannot in any way be
> reconciled with normal verbal usage . . . these spatial words, for the
> blind, have nothing to do with space.[18]

Von Senden, of course, means by 'space', 'space as *we* conceive it'.
It might, today, be suggested that the subject and observer do *not*
speak the same language. Just as two people with different concepts of
'poor' will use the same word but not (in this context) 'speak the same
language' so psychologist/observer and patient/subject will use the
same word space, and not speak the same language. Moreover,
although two people discussing poverty could translate from one
language to another when they (or if they) realised their conceptual
differences, the psychologist and patient might not be able to do this.
It would perhaps be impossible for the psychologist to imagine a

non-visual concept of space or of shape and it would certainly be impossible for the patient to have a visual one. After a successful operation she might, in due course, acquire visual concepts and then could 'speak the same language' as the psychologist.

Von Senden's observations give very strong support to the view that we build our concepts from sense experiences. These experiences are essential; we must have them in order to begin to develop a theory about the nature of the world, in effect to construct the world, *our* world. Yet at the same time we believe that this world exists independently of us and of our perceiving it. Having acquired concepts we then have certain expectations so that when, on any particular occasion, we have particular sense experiences, we *believe that* further investigation will give us certain specific further experiences. For example, when seeing a glass of water (or what we take to be a glass of water) we are directly aware of something colourless and transparent, but we at once assume that we *know* of many other properties which *would be* exhibited, in appropriate circumstances. The *fact* that there is a glass of water (a fact taken *as* a fact because we see the glass of water) embodies our confident belief that if it were tilted the contents would flow, if it were picked up it would be felt to have a certain weight, if we drank it our thirst would be quenched and we should also experience a characteristic neutral taste, if the glass were to be dropped it would probably break and the liquid splash out, and so on.

Our *concept* of water and of all materials and objects develops from an instinctive belief that there is a regular association of properties. This belief generates expectations, that we presume on in our recognition of, say, water, and in our identification of it *as* water. We have this type of expectation in common with many animals (though, see note 5, it does not follow that their classifying scheme is the same as ours). We and they conduct our lives on the assumption that the familiar objects and materials we recognise around us not only have the properties that we directly perceive on any given occasion, but have all the other properties presupposed by the concept.[19]

Of course neither we nor they are always right; our observations can and do mislead us from time to time. A friend of mine was out on a long walk with his dog and noticed that the animal was thirsty. All he had was a bottle of lemonade which he poured into a stony cavity. The dog rushed forward eagerly to drink, took one gulp and then backed away barking; his expectation of water had been disappointed though I trust he found the lemonade not too bad a substitute in due course.

One major difference between humans and other animals is that the latter are less flexible and have more difficulty in adjusting to the unexpected. *We* seem to be able to learn more from experience, in that experience itself teaches us not to be over-confident; so we do, at least to some extent, accommodate 'surprises'. For example, we might find that the colourless transparent liquid we took for water was gin – a pleasant surprise perhaps. But the reason we decide that it *is* gin will be in virtue of our possession of another concept.

In the previous chapter it was stated that our concepts of human beings depended on the 'laws' or regularities taught by experience. Although each individual is unique we do have some general concepts that lead us to believe that almost any person will react in certain expected ways in appropriate situations. Shakespeare portrays Shylock protesting at the treatment of Jews on the grounds that they are as much human beings as are Christians. And why? Because, like Christians, they satisfy the concept of a person:

> I am a Jew. Hath not a Jew eyes? Hath not a Jew hands, organs, dimensions, senses, affections, passions? fed with the same food, hurt with the same weapons, subject to the same diseases, healed by the same means, warmed and cooled by the same winter and summer as a Christian is? if you prick us, do we not bleed? if you tickle us, do we not laugh? if you poison us, do we not die? and if you wrong us, shall we not revenge?[20]

Just as our assessment of individuals is very dependent on concepts formed through personal experience, so assessment of human societies is much affected by the concepts developed through experience of our own society. These concepts, like all concepts, arise by classifying events and customs. We classify materials as elements, compounds and mixtures; we classify living creatures as mammals, fishes, reptiles, birds and amphibia; we classify societies as tribal, feudal and industrial and people within our industrial society as members of social groups A, B, C, D and E. Such classifications are not, of course, made instinctively, they are imposed on us because they are part of our cultural and linguistic background – the terms are part of our language. Any classification involves picking out features that provide criteria for the classification and the features chosen are those that are of interest and concern to the members of the society – hence, as we saw, Eskimos classify snows whereas Britons do not. In general we find it easy to accept new classifications based on an interest in (for us), strange physical features but we find it difficult to accept classification based on different values.

Early Western explorers were shocked at the nakedness of people living in warm countries; their societies were classified as 'primitive' or 'heathen' and these classifications reflected Western values and were not merely descriptive. Today we have a morally neutral attitude to this feature of some alien societies though we do not invariably take a neutral moral attitude to alien behaviour and practices, even behaviour that the society itself condones or positively encourages, and even to practices that are fundamentally characteristic of that society. But rejection of a custom or way of life cannot be on the grounds that it is 'unnatural'. Much behaviour that we regard as highly immoral, such as slavery, cannibalism, torture and racialism, is 'natural' in other societies and, as was pointed out in Chapter 1, it is necessary to understand the indigenous values and morality before rejecting them and *a fortiori* before trying to change them.

As we learn more about the world we learn to allow for the unexpected and unusual physical event and we may also learn to make some allowance for what we take as alien values. The process of learning has to be relatively slow for, if the world presents too many surprises, that is if we cannot establish some regularity or pattern in the course of events, we would be completely disorientated. We could not form stable concepts and the chaos of experience without significance would be akin to madness. It is only *after* a relatively stable framework has been set up that we can appreciate surprises as surprises and can then use them to modify the framework. Through such modification our concepts become richer and more flexible for to learn more about the world is not only to grasp more concepts but also to extend the richness and depth of current concepts; that depth can give increased flexibility without loss of clarity.

Factual descriptions of any phenomenon necessarily reflect the concepts and expectations that we have but since no description can ever be complete the totality of facts available cannot be given; just some of the facts are selected for the description. We select what we think is relevant, important and interesting but there is always a danger that some facts that, later, will be shown to have great significance will be overlooked and omitted. For example, Darwin had noticed that there were many varieties of finch in the Galapagos

Islands but it was not until the Governor happened to remark that he could tell which island any particular variety of bird came from that Darwin realised the potential significance of the fact and that he should have recorded not only the differences between the finches but also *where* he had observed each variety. He was not able to make good the lack of information for the *Beagle* had to sail on. Another instance of an important fact ignored was revealed when a general practitioner published his observations of the frequent occurrence of birth deformities after mothers had suffered German measles in early pregnancy. The *fact* had been available for decades[21] but the significance had not been appreciated, and therefore it had been ignored. Since the Second World War there has been extensive inquiry to explain the increase in lung cancer; it was thought that this might be due to increase in motor traffic and exhaust fumes, to the tar on roads or to some new material used in buildings. The fact that smoking had increased was ignored at first so when the inquiry began there was no investigation to find out whether the increase in lung cancer was connected to increased smoking.

This risk, of not recognising, of ignoring or missing the significance of facts, cannot be avoided for some selection must be made and some guiding assumptions have to be accepted. We must rely on something in order to ask questions; if nothing was accepted nothing could be understood and no questions could be asked. But once questions are asked, the search for answers may lead to a modification of the original assumptions and concepts that prompted the questions. We seek explanations in order to satisfy curiosity and gain knowledge and understanding, and in doing this we also come to criticise current concepts and the current body of knowledge. Modifications of our concepts entail modifications of our facts, modifications of our view of the world. If we were to maintain that facts were rigid we would not be able to make progress beyond the confines of the framework of our current concepts and the overall conceptual scheme which they form and in which they are set.

Just because it is essential for us to have such a framework, and to rely on it, it is very difficult for us to conceive of anything radically different. It needs exceptional imaginative powers to break free from the presuppositions of the system through which we learned how to organise our experiences, and to envisage an alternative that demands fundamental change. Indeed it is psychologically difficult to accept a new conceptual scheme even after it has been formulated, for such acceptance involves jettisoning what we regard as firm facts. We can appreciate this when we consider the hostile reception accorded to certain scientific theories that entailed profound modification to the established view of the nature of the world and of human beings; for example the reaction to Galileo's version of the Copernican cosmos which was referred to in Chapter 1.

Galileo's new physics made the Copernican sun-centred solar systems a physical possibility (as opposed to an elegant hypothesis for predicting the motions of the heavenly bodies) and his telescopic observations indicated that the new cosmology might be a truer description than that offered by the Aristotelian/Ptolemaic account. But it entailed a revolutionary change not only in the conception of the physical universe but also in a number of other concepts – it required a fundamentally new concept of the place of human beings in nature, of their relation to each other and to God. It is too simplistic to explain the resistance as being due to religious bigotry and the tyranny of the Roman Catholic Church. There was, of course, some religious prejudice but probably much more important was the pervasive human desire to maintain the *status quo*, intellectually, socially and spiritually. The old cosmology was not only incorporated into Christian doctrine, it was supported by and in turn supported contemporary physical science, it supported and was supported by the feudal hierarchical social structure, it inspired literature and art. The Aristotelian vision was of a finite, calm and ordered universe: above the earth heavenly harmony (the music of the spheres), a harmony to which human beings could aspire; all this had to be abandoned, or consigned to the realm of myth. Small wonder that the new concept of a vast, potentially infinite, universe, in which the earth was an insignificant planet, was unwelcome. Darwin's theory of evolution and Freud's theory of the unconscious mind met with analogous hostility. Theories involving less fundamental changes in outlook, such as the oxygen theory of combustion, the germ theory of disease, the Mendelian theory of inheritance, were also resisted but with less strength – they caused less disruption.

Any change in concepts must modify our view of the world, and a major change in concepts produces a major change in our view of the world. Certain philosophers of science[22] have argued that when our concepts change the world itself changes, not just our human beliefs about its nature. Such a position is not supported here, and such statements as 'The world changes as our concepts change' are acceptable only if intended in a metaphorical sense meaning 'The world as we see and understand it changes as our concepts change'. In this text the assumption is that, within the confines of our human perceptual and reasoning powers, there is an objectively true description of the world and that we can legitimately hope to attain this truth and have empirical knowledge, that is knowledge of the world. Moreover it is here maintained that it is not irrational to claim that human beings already have some empirical knowledge and will acquire more.[23] As we know, there have been colossal mistakes; this is not surprising, for if we believe there is such a thing as objective empirical truth we must accept that there is also objective empirical falsity, and we must concede that mistakes can be made. Medieval philosophers thought that they *knew* that the earth was at the centre of the universe and now we say that they were *wrong*; we do *not* say that the universe was different then. Undoubtedly some of our current firmly established beliefs about the world will be shown to be false; it will turn out that we did not know what we thought we knew. But then we shall modify our view of the world – the world itself will not alter. It is necessary to stress that just because there must always be the possibility of mistaken beliefs it does not follow that we are never justified in claiming knowledge. *A fortiori* it does not follow that there is no such thing as objective truth about the world, an objective truth which human beings can not only understand but can also aspire to.

This is not to say that, *at the time of change*, the evidence is unequivocal so that anyone who is rational and not prejudiced against change must accept the new theory as a true account and concede that its predecessor was erroneous. In the early seventeenth century Cardinal Bellarmine had rational *arguments* to oppose the Galilean version of Copernicus's theory. He was prepared to allow, as Jesuit astronomers, along with other astronomers, had allowed since 1543, that the theory could predict the positions of the heavenly bodies (see above, p. 22) but he thought that the Bible itself provided indisputable support for the theory that the earth was at the centre of the cosmos – this theory was *true*. As Rorty says:

> What determines that Scripture is *not* an excellent source of evidence for the way the heavens are set up? . . . There were attempts to limit . . . the scope of Scripture (and thus of the church) – the opposite reaction to Bellarmine's own attempt to limit the scope of Copernicus. So the question about whether Bellarmine

(and perforce, Galileo's defenders) was bringing in extraneous 'unscientific' considerations seems to be a question about whether there is some antecedent way of determining the relevance of one statement to another.[24]

. . .

We are the heirs of three hundred years of rhetoric about the importance of distinguishing sharply between science and religion, science and politics, science and art, science and philosophy, and so on. . . . But to proclaim our loyalty to these distinctions is not to say that there are 'objective' and 'rational' standards for adopting them. Galileo, so to speak, won the argument and we all stand on the common ground of the 'grid' of relevance and irrelevance which 'modern philosophy' developed as a consequence of that victory.[25]

Of course our *grasp* of reality depends on our thoughts and on our interpretations of sense experiences; the first chapters of this book stress the dependence of the world *as we perceive it* on our concepts, theories and languages. In a very important way we build the facts of the world about us but, also in a very important way, those facts are not entirely dependent on what we construct. As Trigg says:

Everyone accepts that scientific theories change our conception of reality, but the question is whether there is a sense in which they actually change reality. . . . The world as conceived by scientists does change, but does that mean the world itself changes? . . . Can they [scientific theories] be measured against anything external to themselves, and therefore at least in principle be judged correct or mistaken, true or false? The alternative is that we are left with a succession of different theories, or conceptions of the world, with no means of determining which is better than others.[26]

If we insist that our concept of the world *is* the world we are, in effect, asserting that our concepts cannot be inappropriate and *must* relate to the world. Then questions of truth and falsity become irrelevant. Yet is this is so, why should we bother to inquire further, for whatever we *think* is the case *is* the case?

In the next chapter we shall see how Scheffler deals with the problem presented by the fact that observation must be guided by theory and yet must also be tested by theory – the paradox of common observation. Here we are concerned with the broader problem posed by the fact that we must make sense of the world by relating what we perceive to some overall conceptual scheme, or related set of schemes – a paradigm or one of a set of compatible paradigms. The paradigm will provide the framework for what is established knowledge; in the Middle Ages the cosmological paradigm was that of Aristotle and Ptolemy and as we have seen, this was itself subsumed by the Christian Church into a more comprehensive paradigm incorporating religious beliefs – taken at the time to be largely empirical beliefs. Today, due to the realisation that even Newtonian physics was not perfect, we are less dogmatic in claiming certainty for science but we rely on our current paradigms, for example our physical theory about the subatomic structure of matter and the existence of fields of force; these are advocated and upheld by the scientific establishment that has replaced the Christian Church as authority on empirical matters.

Those who believe that the world itself depends on our concepts (see note 22), will also support the view that empirical knowledge is what is accepted by the authorities (or establishment) of the day.[27] The notion of knowledge is related not so much to individuals *qua* individuals but to what society takes to be knowledge.

> The sociology of knowledge tells us that knowledge is a human construct, grounded in the facts of our social life. It denies traditional definitions of knowledge which insist that knowledge must involve *true* belief, and reflect reality. For the sociologist of knowledge, reality is what is reflected by the beliefs of society.[28]

In a sense this is correct – the Aristotelian/Ptolemaic account of the cosmos *was* medieval knowledge – but it is, for that very reason, highly misleading. For, like the view that each person's concepts of the world *are* the world, it rules out the notion of mistaken concepts. What the social group takes to be the nature of the world *is* the nature of the world. Thus according to this thesis the medieval establishment had knowledge of its world and we have knowledge of our world. As before the contrast between knowledge and belief disappears and the notion of objective truth becomes redundant. As Trigg points out:

> Far from looking at the content of knowledge and trying to assess its worth through reason, the sociologist of knowledge will refuse to abstract what is known from its social setting. Any remaining

contrast between knowledge and belief would have to be between
what is socially accepted and what is individually held. . . . The
social sources of authority and the institutional basis for what is
counted knowledge are the focus of sociological concern. . . . The
judgements of the individual gain their sense from a public
framework. Reality is the product of social agreement.[29]

Trigg says that this position is self-defeating for it cannot justify itself
except by appeal to objective truth that is *not* dependent on the beliefs
of the social group.

each explanation turns out not to be valid at all, but only what a
particular group holds valid. . . . the explanation may be accepted
because of social conditions and group interests which have
nothing to do with the explanation. There can be no reason why
any such sociology should be accepted by other groups who may
have their own interests. . . .

Unless sociologists are prepared to assert certain things as *true*
and to maintain that certain types of social conditions *are* causes,
there is no reason why non-sociologists should accept anything
they say. The trouble is that, if sociologists of knowledge do this,
they undermine their own thesis.[30]

There must be some paradigm, or set of paradigms; we make our
facts more firm and our concepts more stable by making them
interdependent in a paradigm setting. Our need to believe that
perception gives us knowledge and that our investigations into the
world are not fruitless tempts us to believe that our paradigms are
right and hence we may warm to a philosophy that supports the view
that the way we (or our social group) *think* the world is, *must be* the way
it is. But here lies danger, for if we go along with such a thesis we
jettison the concept of objective truth along with the concept of
objective reality and then there is no possibility of objective
knowledge. It is our desire for certainty that leads us astray; we have
to accept that if we espouse a philosophy that makes certainty certain
then the baby of truth goes out with the bath-water of scepticism.

3

Facts and Theories

In the previous chapter it was stated that all empirical facts, even the most simple, those seemingly directly 'given', depend on the interpretation of sensation by means of concepts and that concepts are the basis of knowledge. Our concepts of familiar objects and materials *and* our concepts of people and social customs depend on our early experiences and develop from what were called 'instinctive beliefs', based on a fundamental human (probably animal) tendency to trust in and to expect a regular association of properties and qualities. Therefore, on seeing a familiar object, such as a glass of water, we all assume that it has properties that are not currently perceived. Now instead of relating our instinctive beliefs to concepts, we could say that they emerge as a result of spontaneous theories as to the nature of what we perceive. Such theories support our concepts and give rise to our expectations (predictions) about the world. It may seem downright misleading to assert that such expectations involve the acceptance of theories, for we take the properties of well-known and familiar objects and materials to be matters of fact and we *contrast* fact with theory. This is because we do not generally reflect on the basis of our knowledge of familiar facts and certainly do not formulate explicitly either our expectations or the theories that support them. Life is too short to spell out our theories and our assumptions about the nature of the world, let alone to justify them.

We can appreciate the dependence of fact on theory more readily if we consider facts that are not so familiar as the facts about everyday objects such as a glass of water. For example, it does not seem strange to say that the recognition of rod-shaped micro-organisms as carriers of tuberculosis depends on the acceptance of a theory that these micro-organisms attack human tissue; neither does it seem strange to say that the identification of a glowing patch in the night sky as a comet depends on accepting an astronomical theory. Just because such phenomena are much less familiar, the role of theory as establishing their nature as a matter of *fact* is much more easily appreciated. Consideration of such facts can confirm the assertion made above that our expectations about the properties of all objects and materials, familiar and unfamiliar, are based on theories as to the regular association of groups of properties. It is just that if a theory is familiar and well-established and *a fortiori* if it has arisen spontaneously, it is not recognised as a theory, it is taken to be a matter of fact. The 'theories' that water flows, that fire feels hot, that bread nourishes and so on, are taken as facts and are built into our concepts of water, fire and bread.

Likewise theories about the 'normal' behaviour of people are taken as facts; some of these were given by Shylock (see previous chapter p. 19) and they are part of our concept of a person. Thus we accept without hesitation, the theory that if people cannot get food they will feel hungry so that 'Starvation causes hunger' is not held to be a theory but a statement of fact. By contrast we may have doubts about the theory (we may even call it a 'conjecture') that neuroses develop as a result of unpleasant childhood experiences being repressed. Just because there are doubts, or reservations, we do not take such an account of neuroses as being an unequivocal factual account. It is, in the view of many, at best a tentative fact, nowhere near as firm as the fact that starvation causes hunger. Accordingly our facts, what we take to *be* facts, depend on our assessment of the reliability of the theories that suggest or educe the facts. Those who support a Freudian theory of neuroses will judge the statement 'Repression causes neuroses' to be a statement of fact. Those who are neutral will judge it to be a possibly factual statement. Those who oppose the Freudian view will dismiss the statement as, at best, 'mere conjecture', at worst as false and misleading.

Here we are not concerned to argue the case for or against this or any other theory; we are using it as an example to show that facts are regarded *as* facts only because the theory or theories supporting them are accepted. Moreover a fact is taken as 'firm' and is (mistakenly) thought to be theory-free, when its educing theory is so familiar and so acceptable that it is ignored. Such a theory has become 'transparent' for we are aware only of the fact or facts, it educes. As Whewell[1] said:

> A Fact is a combination of our Thoughts with Things in so
> complete agreement that we do not regard them as separate . . .
> these Thoughts are so familiar, that we have the Fact in our mind
> as a simple Thing without attending to the Thought which it
> involves.[2]

and

> In a Fact, the Ideas are applied so readily and familiarly and
> incorporated with the sensations so entirely, that we do not see
> *them*, we see *through* them . . . thus a true Theory is a Fact, a Fact is a
> familiar theory.[3]

So even our most familiar facts rest on concepts *and* presuppose
theories, albeit formulated without much conscious thought. *All facts
are theory-laden.* This applies as much to particular statements of fact as
to generalisations. It follows that established theories support not
only facts but also candidates for consideration as facts and as factual
statements. Thus *any* empirical statement must be supported by
theory and, since it can only be judged true or false if it is understood,
it follows that the supporting theory must be understood. Statements
such as 'It is raining', 'The man committed suicide by jumping from
the top of the Hilton', 'The spectrum shows that helium is present' are
significant only to those who have grasped the relevant concepts and
have understood the relevant theories. Then theories can become
established as facts; as Whewell said, a true theory *is* a fact.

Since facts and theories are so intimately related and since both are
involved in any description and in any explanation it might seem that
facts cannot be used to test a theory. For if all facts are theory-laden
then explanatory facts must be laden with theory and cannot but be
compatible with the theory. They are bound to support it and so
cannot be the basis of an independent objective test. This seems to
present an insurmountable difficulty: there can be no observation
that is not 'influenced' by theory, no observation that is theory-free.
The apparent *impasse*, the paradox of common observation, is
expressed by Scheffler:

> There seems to be good psychological reason to suppose that

observation is not at all a bare apprehension of pure sense content, but rather an active process in which we anticipate, interpret and structure in advance what is to be seen. . . . But if observation is never conceptually neutral, if it cannot occur without expectation and schematization, then stripping away all interpretation leaves nothing at all. And if nothing at all remains what is there to provide an independent check on interpretation?[4]

However, as Scheffler goes on to remark, the difficulty can be avoided because what is observed, and the empirical descriptions given, can be independent of any theory suggested to *explain* what is observed. Simple observation and the factual (empirical) descriptions given are indeed theory-laden but they do not have to be laden with the theory suggested to explain them, the theory under test.

For example it is possible to give a description of a man writhing on the floor that is not dependent on any explanatory theory: how he moves, the positions he takes, his facial expressions are facts that depend on theories, admittedly, but basic theories about physical objects and people, not theories *about* the behaviour observed. Those basic theories relate to the direct description; but they do suggest that what is observed in this case is unusual in that general expectations arising from those theories do not include convulsive writhing on the floor. Nevertheless the behaviour is not so strange that it is regarded as illusory, not a matter of fact at all.[5] For this reason explanations are sought and they will depend on *other* theories that educe new (explanatory) facts. The current explanation depends on the theory that there can be malfunctions of the brain that can cause convulsions from time to time. Those who suffer from this malfunction are called epileptics and the periodic convulsions are called epileptic fits. These are the new facts, dependent on the explanatory theory. They may well modify the assessment of the behaviour and *can* be used to describe what happens, but they do not *have* to be used. Moreover the more basic account of the movements holds good whether or not the explanatory theory is accepted and whether or not it comes to be superseded. Thus the *explicandum* will remain whatever the fate of the *explicans*.

Consider a theory explaining the behaviour of a man writhing on the floor which suggests that he is possessed by devils. If this theory is accepted then, for those who do accept it, it is a *fact* that there are devils who have the power (perhaps along with other powers) to produce convulsions from time to time in certain people, or at least in that particular man; those devils are responsible for his convulsions. On the other hand if the theory of brain malfunction is accepted and therefore thought to provide the correct explanation then it is a *fact* (see above) that he is an epileptic. But it does not matter which theory is accepted, or even if neither theory is accepted, the simple and most basic fact that the man is writhing on the floor will remain; it will be unaffected and the basic description need not change. In this context we are not interested in which theory may be correct, we are concerned to show that no explanatory theory can disturb the basic facts it purports to explain. These are independent of any explanatory theory and will therefore be neutral between competing explanatory theories. This is not to say that an explanatory theory has no influence, on the contrary it usually modifies the assessment of the facts and broadens our concepts. Quine points out that to understand events described in familiar terms, for example a man writhing on the floor, we need to appeal to higher-level explanatory theories:

> If we improve our understanding of ordinary talk of physical things, it will not be by reducing that talk to a more familiar idiom; there is none. It will be by clarifying the connections, causal or otherwise, between ordinary talk of physical and various further matters which in turn we grasp with help of ordinary talk of physical things.[6]

The 'ordinary talk of physical things' is the account in the familiar terms of the basic theory and this must be related to explanatory theories, 'further matters'.

So although basic facts, *explicanda*, can be described quite independently of any explanatory theory they *are* influenced; for example our view of a man writhing on the floor is bound to be affected by the explanation be it in terms of possession by devils or of epilepsy. Thus because two competing explanatory theories will influence the facts differently it can indeed be difficult to compare them and this is the ground for the paradox of common observation that has been discussed. Our discussion shows, however, that it is always possible to have a theory-neutral description though there can be psychological difficulty in restricting ourselves to it.

It is less difficult to provide neutral descriptions (free from explanatory theory) of physical objects and events than of human behaviour and hence it is easier to compare competing explanatory theories in the fields of the physical sciences and to make assessments that are objective and independent. Comparison of competing explanatory theories in the human sciences (with the possible exception of medicine) tends to be much less straightforward. There are several reasons why this is so and, as will be seen, these are related to the difficulties we have in coming to agreement about explanations in the human sciences that were discussed in Chapters 1 and 2. Firstly, even at the level of 'ordinary talk of physical things' the facts of human behaviour are less 'firm' and so are more readily modified; secondly (and this will be discussed at greater length in Chapter 14), there is almost always an evaluative element in any description of human beings and human societies so that the distinction between explanation and description is blurred; thirdly, explanations may not be in direct competition, they may be complementary, so that their different influences on the basic facts are also complementary and the problem may be not to decide which theory is correct but which is more illuminating in a given situation. Fourthly, any explanation offered can affect the behaviour of those whose behaviour the theory was devised to explain, that is the theory (and any other explanatory theory) has a direct effect on subsequent behaviour and hence may be self-defeating or self-justifying.

Let us consider two theories that have been offered as explanations of the discontent shown by people who have been moved from inner-city slums to new housing estates. George Orwell suggested that the discontent was due to:

> the bleak atmosphere of the estates. They are built 'in a ruthlessly inhuman manner' and are 'soulless', with, for example, 'dismal sham-Tudor pubs'. The estates are replete with petty restrictions on such things as the decoration of houses and the keeping of pets, restrictions which witness the fact that 'owing to the peculiar temper of our time', in order to secure slum-dwellers decent housing 'it is also considered necessary to rob them of the last vestiges of their liberty'.[7]

Two sociologists, Young and Wilmott, offer a different explanation:

> Much of their book is devoted to establishing the nature and functions of the strongly mother-centred extended kinship

structure of the slums. They argue that a place in such a kinship structure provides one not only with the friendship of its other members, but also with a framework for meeting a wider set of people who are not members of the family. .... Young and Wilmot hypothesise that the key change in moving from the slums to the estates is the break-up of the mother-centred kinship structure. This produces loneliness amongst the estate dwellers, both directly through the loss of family and friends and indirectly through the loss of the framework for meeting people.[8]

These two explanatory theories are not incompatible; they may be thought to be complementary, both sets of factors contributing to the discontent, though one may be inclined to wonder whether the inhabitants of the estates would be so much offended by mock-Tudor pubs as was Orwell! But another explanation was offered by the inhabitants themselves: *they* said that they were dissatisfied because the houses were poorly built and uncomfortable to live in.[9] Now Orwell would agree with Young and Wilmott that the physical conditions, housing and facilities, were much better than those of the slums. All of them would have concluded that the inhabitants were rationalising deeper grievances – Orwell states what he thinks these were, Young and Wilmot suggest others.

Again, we are not so much concerned with which explanation is correct, or whether the two explanations offered by observers are or are not complementary, or whether one is more important than the other, or whether the inhabitants were rationalising or whether they were offering a third, genuine, explanatory factor. Here we are concerned to show that each explanatory theory not only suggests new facts but also has a very marked effect on the significance of the facts explained, that is on what we take to be the nature of the discontent itself. This is firstly because the concept of discontent is somewhat vague and so is liable to modification by any explanation proposed. Secondly, 'discontent' is value-loaded as well as theory-loaded, and the value element is intimately involved with any explanation. For most readers of this book discontent with soulless surroundings and petty restrictions and loss of family relationships will probably seem more 'justified' than discontent with the nature of a pub. Thirdly, if the explanatory theories are held not to be in direct competition but to be complementary, the fact of discontent becomes even more complex. Lastly, we have to acknowledge that the view the inhabitants take of their situation will itself alter the nature of their discontent, and if in addition they become aware of alternative explanations offered by others, their behaviour and attitude are also likely to be altered.

However, even within the human sciences we can make some distinction between theories (and therefore facts) at different levels. Hence, when considering explanations we can propose a hierarchy of facts and theories:

(1) Basic facts about physical objects and people; these are educed by fundamental theories arrived at spontaneously. They provide the hard data of physical science and are part of the 'ordinary talk of physical things'.

(2) Facts that are educed by higher-level theories to explain common-sense facts; these explanatory theories and facts do not emerge spontaneously but as a result of active thought. Here we have the beginning of what may be called 'scientific facts', facts that most of us learn.

(3) More sophisticated scientific facts that are educed by theories explaining the lower-level scientific facts.

At the third stage, if not before, we reach the point where 'the facts' are quite obviously dependent on theories and at this stage facts are generally considered to be less firm, because more sophisticated theories are generally less well established. So in both the physical and human sciences there will come a level of explanation where there is a difference of opinion as to the reliability of the explanatory theory. Therefore there will be a difference of opinion as to whether the 'facts' offered by a theory are indeed facts. As we have seen, such differences appear more commonly in the fields of the human sciences.

Let us consider two sets of examples: one from physical science and one from a human science:

(1) (i) Certain solids melt in the sun.
    (ii) Melting is explained as being due to a rise in temperature.
    (iii) The rise in temperature is explained by the molecular theory.

It is worth noting that the relatively sophisticated molecular theory has explanatory power because it presupposes that micro-objects (molecules) can behave, at least to a certain extent, in ways analogous to the way ordinary physical objects (for example billiard balls) behave.

(2) (i)   House prices rise.
    (ii)   The rise is explained as being due to increase in demand.
    (iii) The increase in demand is explained by:
           (a) the theory that people judge property to offer good protection from inflation;
           (b) the theory that people avoid capital gains tax on their homes.

Here (a) and (b) are not incompatible, but complementary. Again we may note that these two theories have explanatory power because both presuppose that people wish to protect and increase their assets.

The hierarchy of facts and theories shows both the continuity and the distinction between common-sense observation and theories and scientific observation and theories. Common-sense facts, embodied in 'ordinary talk of physical things' give us the hard data of our world even though we can and do find that sometimes they have to be modified and occasionally rejected. We saw in Chapter 1 that the common-sense account of the daily passage of the sun across the sky came to be rejected some centuries ago and more recently the common-sense maxim 'Spare the rod and spoil the child' has come to be substantially qualified. All the same, as indicated above, we rely on common-sense assumptions and theories to help our understanding of the 'various further matters' that are the substance of the explanatory theories.

What is taken to be 'ordinary talk of physical things' will differ for different persons for what each individual takes to be a matter of common-sense fact will depend on her background knowledge. This appears in the examples of the two sets of hierarchies already given, and some more examples will show how the distinction, though undeniable, cannot be clear-cut;

(1) Water is essential for life.
(2) Water is composed of hydrogen and oxygen.
(3) Water has a pH of 7.
(4) Evaporation of sweat keeps us cool.
(5) High humidity makes us uncomfortable.

(6) Smoking increases the risk of dying of lung cancer.
(7) As prices rise demand falls.
(8) Alcohol affects judgement.
(9) Alcohol reduces brain activity.
(10) Alcohol is a cause of crime.

These ten facts,[10] and let us, for the purpose of discussion, allow them to *be* facts, depend on various theories. It is our view of the theory (as being one of common sense or of science) that determines our view of the status of the fact. Some, for example (1) and (8), are unequivocally common-sense facts for any adult; others, for example (3) and (9), are unequivocally facts of science. Others, for example (2) and (7), will be rated by some as scientific and by others as facts of common sense. The distinction between common-sense facts and facts of the human sciences is even more blurred than that between common-sense facts and facts of the physical sciences. But, though there will be differences of opinion as to the status of the facts listed above, it does not follow that there are no criteria to guide judgement as to the status of a given factual (or possible factual) statement.

There are two criteria: firstly level of theory; this has been discussed above, it affects the familiarity of the fact. Secondly, the precision of expression of the fact. Precision may involve the use of technical terms such as pH, understanding of which involves understanding of higher-level theories; precision may also involve the use of mathematical relations and, lastly, it may involve deliberate restriction of the description, limiting it to an account of what is (in the light of the theory) considered relevant.

Facts at all levels must be expressed in language, and a language is meaningful because it arises from concepts and theories. Different concepts and theories express different facts and give rise to differences of language. Hence the same words, for example 'atom', 'poverty' and 'religion' have different meanings when supported by different concepts. This is why individuals with different concepts can fail to communicate; they do not appreciate that though they are using the same words, in the context of the explanation they are not speaking the same language. We need to consider the relationship between facts and language.

4

Facts and Language

The world is *my* world: this is manifest in the fact that the limits of *language* (of that language which alone I understand) mean the limits of *my* world.[1]

'What we cannot speak about we must pass over in silence.'[2] From our consideration of the dependence of empirical facts on concepts and theories we have come to appreciate that all observation involves inference and that we are not the passive recipients of hard data supplied by the objects and events in the world: we interpret our sensations in terms of observed objects and events because we are the kind of creatures that we are. We speak of *facts*, and now we are in a position to tackle a very difficult question, namely 'What are facts?'. The immediate answer suggested by common sense is 'Facts are the situations we find in the world'. It was expressed by Wittgenstein as: 'What is the case – a fact – is the existence of states of affairs.'[3]

But this sort of answer is not very satisfying, for one is then prompted to ask *how* it is that facts represent the world? *Can* they be an objective description of 'what is the case'? For we now know that facts are theory-laden, laden with *our* theories; they do not come to us as 'ready-mades', they are, at least partly, made by us. How can we come to grips with the problem of the nature of facts, of what kind of entities they are? We can 'point to' particular objects and events but objects and events are not facts; they exist during a particular time and at a particular place (or at particular places) and that is why, if we are in the right place at the right time, we can (at least in principle) 'point to' them. However, the *facts* corresponding to objects and events, for example the fact that it is raining in Exeter on the morning of 11 October 1986, are timeless; and though that fact *refers to* a particular place (Exeter) it is not itself *located* in any particular place. In this respect an empirical fact[4] is like a mathematical or logical fact such as '2 + 2 = 4'.

The distinction between facts and 'what goes on in the world' is even clearer if we turn to empirical generalisations such as 'All sea-water is salty', or 'Heavy smokers are more likely to die of lung cancer than are non-smokers'. For though the particular instances that support and justify empirical generalisations occur at specific times and in specific places[5] the generalisations themselves do not. Empirical facts are true accounts of objects and events and they are expressed in language – we can say that they are made concrete in certain kinds of sentences, namely empirical *statements*.

Statements are a special sort of sentence, they are sentences in the indicative mood. Thus 'The cat is on the mat', 'The cat was on the mat', 'The cat will be on the mat', 'The mat is under the cat', 'The mat was under the cat' and so on are all statements. By contrast 'The cat might be on the mat', 'Is the cat on the mat?', 'Oh that the cat were on the mat!', and 'Put the cat on the mat', though all meaningful sentences, are not statements. The feature common to all statements is that they can be asserted and, if they are asserted (as opposed to being used as examples of sentences, see above) then they have a truth value – that is they must either be true or false. If they are true they assert a fact.

Now any statement, such as 'The cat is on the mat', can be asserted in many languages: French, German, Japanese, and so on – any language which copes with the relevant concepts. Each of the statements is an expression of the *same proposition*. In effect propositions are what statements state and a statement can be regarded as a concretised proposition. (Hence, in certain contexts, the term 'proposition' has the same sense as the term 'statement'.) Different statements in the same language can also express the same proposition. Thus 'Joan is taller than Jane' expresses the same proposition as 'Jane is shorter than Joan', and 'John loves Mary' expresses the same proposition as 'Mary is loved by John'.

Facts, therefore, are related to statements and to propositions but they are not themselves statements or propositions, not even true statements or propositions. Empirical facts 'hover' between events and objects in the world and what we think and say about those events and objects. They emerge from concepts and theories that are themselves prompted by sensory experience, and they link our own sensory and conceptual reaction *to* the world *with* the world. That world, the world we become aware of through perception, is to a large extent 'made' by us; empirical phenomena conform to our ideas because empirical facts are generated by us and are laden with our theories. We structure our sensory experiences, that is the way we make them significant; yet in a very important way the world is independent of our experiences and independent of what we think and say. The facts that we ourselves construct have status as *facts* just because they are *not* to be at the mercy of our theories, our expectations and our beliefs. As O'Connor says:

> a belief (or its expression) must be true in virtue of something external to the belief itself but to which the belief is *in some way* related. Truth then is a relational property and a theory of truth must spell out the nature of the relation.[6]

Only a true statement asserts and embodies a fact, a false statement may be said to assert and embody a fiction. By asserting what we take to be facts, true statements also assert our beliefs about the world. False statements may also assert our beliefs (if we mistakenly think that the statements are true) or they may assert what we want others to believe or what we imaginatively create. This is set out in Table 1.

We see that empirical statements have two functions: they embody and assert facts (or fictions) and they provide descriptions. We can *think* about the two functions separately, but any given empirical statement will have both functions, just as any given object (or sample of matter) will be composed of some substance and will also have some shape. The descriptive function of statements is related to sense perception (or imaginings derived from sense perception), and so links the statement (and its associated proposition) back to the world from whence it originated. The assertive function is related to facts (and fictions) and so links statement and proposition to concepts and theories. For statements consist of words and they are significant because the words are not mere noises or marks on paper but relate to our concepts and theories.

We have seen that concepts change as theories change but that words may not change so that the same word may, at different times, have different meanings. Therefore statements composed of the same

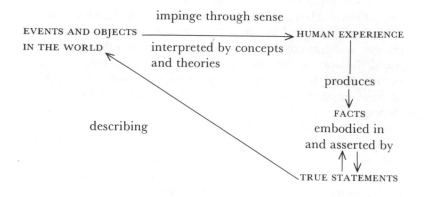

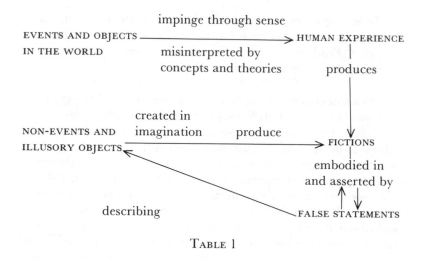

TABLE 1

words may express different facts (or fictions) and may describe different states of affairs. As we saw at the end of Chapter 3, in the appropriate context they are in different languages. For example, the statement 'The sun travels from East to West' has a different meaning for us as compared with its meaning for Shakespeare. Shakespeare and his contemporaries believed that the sun moved round the earth each day. Not a significant alteration in the sense of the statement, after all we certainly speak of the sun travelling across the sky today? But it is! Our concept of the sun's motion is fundamentally different from Shakespeare's. Take Hamlet's letter to Ophelia; today it would not be understood as giving a pledge of constancy:

> Doubt thou the stars are fire;
> Doubt that the sun doth move;
> Doubt truth to be a liar;
> But never doubt I love.[7]

Many words had different meanings and therefore many statements had a different significance in the sixteenth century. Yet we are still enthralled by Shakespeare's poetry because much of its descriptive force is related to concepts that have remained relatively stable. For example our concepts of a person (see Shylock's speech quoted in Chapter 2), of plants and, to a lesser extent, of animals, have much in common with those of the Elizabethans. Nevertheless the fact that much remains the same must not mislead us; even our ordinary talk, with the same words, has a change of tone. Discussions as to whether *The Taming of the Shrew* is a cruel play, whether Desdemona was not ridiculously obtuse, whether Shylock is portrayed as impossibly savage or Lear as absurdly obstinate arise partly because our view of unexceptional behaviour has changed; Shakespeare and his contemporaries had different concepts of wife, husband, Jew, Christian, father and child and therefore a different opinion of their normal behaviour. Thus though we understand and enjoy his works they present problems that were immaterial to the Elizabethans. Obviously these are not so great that appreciation, let alone communication ceases, as Putnam says: 'Meaning . . . is a coarse grid laid over use'.[8]

In his *Philosophical Investigations* Wittgenstein gives an account of language as a dynamic creation that acquires its significance through use; the well-known phrase 'language game' refers to a language having meaning through the rules that guide its use, just as a game is the game that it is in virtue of the rules that guide the play. This stresses the assertive aspect of statements, that they embody facts and our interpretations of the world; our concepts and theories impose rules on sensory experiences that give them significance. Putnam points out that language also functions as a picture or map, so providing descriptions; the two views are complementary:

It is essential to view our theories as a kind of 'map' of the world . . .

if we are to explain how they help us to guide our conduct as they do. But the 'use' theory is *also* right as an account of how language is *understood*. And the insights are not incompatible: a map, after all, is only successful if it corresponds in an appropriate way to a particular part of the earth, or whatever. Talk of use and talk of reference are parts of the total story, just as talk of switch-flipping and talk of electricity flowing through wires are parts of a total story.[9]

In Chapter 2 we saw that language and concepts are interdependent; this can readily be accepted. It is also not difficult to accept that our language will reflect the way we interpret our sense experiences and that therefore it must be intimately involved with the way we think about the world as well as the way we describe it. It follows that empirical facts are never concept-free and, since concepts are based on theories (including instinctive beliefs) facts will incorporate theories – they are theory-laden. They can indeed be said to *be* true theories as Whewell claimed. Moreover, in a very important sense facts exist for us[10] only when expressed in language. Hanson echoes Wittgenstein (see the second quote in the epigraph above) when he says:

We can perceive only what we can express, or can to some extent express. What would seeing the utterly unsayable be like? What would perception of the inexpressible be like? Who will indicate for us the facts that in principle elude statement?[11]

So language does not merely symbolise and describe what we perceive and what we think; to a very large extent it tells us *what* we perceive and, to an even greater extent, it moulds our thoughts. Language is part of our world of empirical facts and contributes to the emergence of our explanations about the events in the world and to new ideas about the nature of the world. The specialised technical terms used by those who investigate the world not only embody newly discovered facts, they also help the process of discovery. Thus:

before the Newton–Leibniz calculus certain events certainly could not be seen as they are seen now; our knowledge of the facts of nature was altered considerably by this powerful formal instrument. And due to the strictures of another recent language, the language of quantum theory, the facts of the microphysical world are now understood in a manner profoundly different from what we should have predicted in 1920.[12]

In the next chapter we shall consider the uses of technical terms and how they can help to give precise and simplified descriptions of certain aspects of the world, but we need to bear in mind that they also give us a different picture of the world and help to provide new explanations of what we perceive.

5

Technical Terms

In Chapter 3 it was stated that though there was no clear-cut distinction between common-sense and scientific facts, there were two criteria to guide assessment of the status of facts, namely level of theory and precision of expression; it was also stated that precision of expression could involve the use of technical terms. These criteria also apply to the distinction between common-sense and scientific explanations, though, as with the distinction between the two types of facts, it cannot be clear-cut. In so far as the first criterion, level of theory, is concerned, there has been discussion in Chapter 3, but the use of technical terms to help precision of expression deserves further consideration.

Technical terms, including mathematical formulae, are not introduced in order to make explanations arcane but in order to make them more accessible by making them less cumbersome and less ambiguous; they are then easier to understand. This seems paradoxical only because the understanding rests on knowledge of scientific and mathematical theories; once these are grasped, and the relevant mathematical techniques mastered, the descriptions and explanations will be clearer and more comprehensible than they would be if they were expressed in ordinary (common-sense) language, that is the ordinary talk of physical things. An example showing how mathematical techniques simplify is the comparison of adding, say, 673 to 598 by using the integers and decimal places (i.e. school addition) with counting individual unit strokes. The task would be wearisomely long and there would be a much greater chance of making a mistake. Yet to do school addition involved learning the manipulation of the number system (at school) and it is not until the technique is mastered that the calculation is easier than counting unit strokes. The notation itself is also important: if Roman numerals and notation were used it would be much more difficult to carry out simple calculations; compare multiplying DCLXIII by DLXXXXVIII with multiplying 673 by 598. When the present notation started to be generally adopted intelligent people of the day had to take lessons in addition, subtraction, multiplication and division – what we call simple arithmetic. Samuel Pepys, for example, studied kindergarten arithmetic when he was Secretary for the Admiralty. It was worth his while to take lessons for once he had acquired the technique his computations would be easier, quicker and more accurate.

Now let us consider the advantage of using a technical term such as pH. If a solution is said to be pH 3 then those who understand the term know not only that the solution is acidic, and quite strongly acidic, but precisely *how* strongly acidic. The term 'pH' is related to the inverse of the concentration of hydrogen ions (positively charged atoms of hydrogen) in the solution; these ions are produced when acids are dissolved. Pure water has a lower concentration of hydrogen ions than any acid, it is neutral at pH 7; alkaline solutions have a lower concentration of hydrogen ions than pure water so they have pH values higher than 7. The details of the theory are not suitable for a philosophy text but even this short account suffices to show the value of pH terminology (pH language) in discussing acidity and alkalinity. Anyone needing to discuss or refer to this from time to time finds it worth studying the theory, just as Pepys found it worth studying arithmetic. By learning appropriate mathematical techniques and by studying explanatory theories we grasp the concepts and are able to use the scientific language that is necessary for precise descriptions and clear explanations; such terms as 'isotope', 'galaxy', 'electron' and 'gene' must be understood in terms of the theories that educe them; this was shown in Chapter 3. If the theories are not understood the technical terms cannot be significant and any explanation must be in ordinary language; it would be verbose and very difficult to follow and it would probably be ambiguous and misleading.

The human sciences have their technical terms: the statement 'High interest rates bring about deflation' uses terms that require some grasp of economic theory if the statement is to be understood. Unfortunately words like 'interest' and 'deflation' are not so obviously technical as 'isotope' or 'gene' and hence people can believe they understand them when they do not. Explanations given by economists and other social scientists are liable not to be ignored as being too technical (this is the fate of many explanations in physical science) but rather to be misunderstood. In addition the technical terms may be misapplied to produce a meaningless, or semi-meaningless jargon. Certain social science terms are particularly vulnerable because they 'carry over' their common-sense meaning in an ill-defined manner. In everyday life many terms are 'open-ended', that is their meaning is not fixed and depends on the context and the user; we have already considered 'poverty' in Chapter 2, other open-ended words are 'hot', 'middle-aged' and 'cheap'. As we have seen, the flexibility can be misleading and can confuse ordinary discussion, but flexibility can also be convenient for too rigid a meaning could be constricting. For example in ordinary conversation it is generally perfectly appropriate to say 'It is raining' rather than 'The current rate of condensation is 0.1 inches per hour'. Admittedly some of the terms used in physical science such as 'work', 'salt' and 'insect' are also used flexibly in ordinary talk but in the language of science they have a precise meaning that makes them unambiguous and therefore not misleading. By contrast some of the terms used in the human sciences have retained their common-sense fuzziness and so explanations using them are also liable to be fuzzy. A vague explanation depending on fuzzy terms and concepts is unsatisfactory, especially if the meanings can be 'adjusted' to account for anything that happens. It might be thought that an explanation of this sort, one that was compatible with any outcome, was the perfect explanation, but this is far from the case. As we shall see in Chapter 9, an explanation that can account for anything that is observed *post hoc* cannot be a source of predictions and cannot be tested by observation; it is irrefutable *in principle* for no observation can show that it is wrong.[1] Thus an explanation that can account for anything that might happen is as useless as one that can account for nothing.

To show how vague terms can devalue an explanation consider a suggested explanation of increases in vandalism in cities: 'Increases of vandalism in cities are due to rises in the number of people unemployed'. If this explanation is correct, or partly correct,[2] then it should be possible to show that as unemployment has increased vandalism has increased and that if unemployment falls then vandalism decreases. Alternatively, or additionally, it should be possible to show that the incidence of vandalism is lower in cities with a proportionately smaller number of unemployed and higher in cities with a proportionately larger number of unemployed. Investigation will not be easy for there will be other factors, apart from unemployment, that will affect the prevalence of vandalism and which may vary at different times and in different place; any results must be carefully assessed. But these results will be completely useless unless there is an unambiguous concept of vandalism and of unemployment. Is 'vandalism' to cover a range of social offences from crimes of violence, such as mugging, to the careless dropping of litter? Or is it to be confined to a range of intermediate activities such as the raiding of telephone boxes and breaking windows? Is 'unemployment' to signify all those without paid work, or just all men without paid work, or men of a certain age, or those who have been unemployed for more than six months, or those who have never had paid work and so on? If these terms are not clear and are not used consistently it is likely to be relatively easy to find data that support the explanation and equally easy to find data that refute it. The same applies to alternative explanations: 'Vandalism has increased because policing is less adequate', or 'Vandalism has increased because we have lost our Victorian values'.

Even when the terms are clear and unambiguous, and are used consistently, it is extremely difficult to establish conclusive refutation of an explanatory theory. Elsewhere[3] I have described the problems involved in testing scientific theories. As we shall see in Chapter 11, there are special difficulties associated with the theories of the human and social sciences but if vague terms are used testing becomes not difficult but impossible.

An analogous distinction between the human sciences on the one hand and physical sciences on the other arises in the use of mathematics. Mathematical descriptions, and explanations involving them, presuppose measurement. Measurement is a distinctive characteristic of the physical sciences and has been a distinctive characteristic of causal explanations in these sciences since the time of Galileo (1564–1642). One of the reasons why progress was so rapid from the seventeenth century onwards was that properties previously thought to be unmeasurable were reassessed and redefined so that they could be measured.[4] For example it was known that as a body fell through the air it descended faster and faster; it was also known that bodies moving horizontally could gain and lose speed –

they were said to be subject to difform motion. So there was a concept of what we call acceleration – it was related to distance travelled. But though it was appreciated that the greater the original height of a body, the faster it would move as it approached the ground, there was no way of measuring the acceleration, or of calculating it. Galileo proposed that acceleration should be treated as increase in velocity (speed in a given direction) in a given *time*, as opposed to increase in velocity over a given distance. It was a reassessment of the concept of acceleration, a reassessment that took him seventeen years to arrive at! So defined, the acceleration of a freely falling body (the acceleration that we today call acceleration due to gravity) can be shown to be approximately constant at 32 ft per second per second; that is for each second of fall the velocity increases by 32 ft per second. Conceived in this way acceleration due to gravity is not only readily measurable, it can (because it is nearly constant) be used as a basis for simple mathematical laws that describe how bodies fall and can predict their position and velocity as they fall. Today we have defined many other properties: viscosity (that is a liquid's resistance to flow), magnetic charge, electrical resistance, sound intensity and so on, so that they can be measured and can feature in mathematical laws.[5]

But qualities that feature in the human sciences have not, in general, been conceptualised in ways that make them measurable. For example our concepts of business efficiency, profitability, work satisfaction and generosity have the vague, diffuse and ambiguous form that concepts of acceleration and viscosity used to have. It is because they are not clearly defined that there can be no accepted and objective way of measuring them – we cannot even begin to suggest a method of measurement until the concept is clear. It could be that our difficulties are analogous to our difficulties with the concept of acceleration and that a reassessment would make such concepts more amenable to measurement. If this were the case then there is a possibility that the human sciences could become more precise and mathematical and could produce explanations that were more objective and therefore more likely to be generally acceptable.

It may be that there is a more fundamental difference between explanations of human behaviour and explanations of the behaviour of inanimate objects. There is no general agreement on this point, but it is probable that there are some irreducible differences between explanations pertaining to ourselves and our fellows, and physical explanations. This has already been referred to in Chapter 1 and will be discussed at greater length in Chapters 10 and 11. But, even if we grant that there are *some* irreducible differences in the two kinds of explanation, it does not follow that explanations of human behaviour have *nothing* in common with explanations of physical events.

In any case, whatever conclusions may be reached on this matter, it remains true that where they can be used, mathematical relations give us more precise and more concise descriptions and explanations than do non-mathematical descriptions and explanations. Yet mathematics can mislead, by presenting an apparent precision that is spurious. This can happen in two ways: firstly the mathematical terms may not be understood by those who use them. Such terms as 'statistical correlation', 'average' and 'mean' are frequently used incorrectly and then the proffered 'explanations' are nothing more than jargon. For example, one hears 'A-level results are *not correlated with* classes of University degrees'. Now those who assert the proposition may intend no more than to suggest that not everyone who has outstanding A-levels gets a first class degree, and that not everyone who just scrapes in with the minimum A-level, gets a third class or fails. This is likely to be true, but it has a negligible bearing on the truth of the proposition denying correlation.

Secondly, mathematics may be applied to concepts that are not susceptible to objective measurement, or not measurable in a way that is universally acceptable – for example 'productivity' and 'intelligence'.[6] Thus, if we say 'She has an IQ of 121', our statement has a spurious precision. It means little more than 'Her intelligence is above the mean' a statement which almost certainly could be justified without any appeal to formal intelligence tests and the resulting numerical IQ. By contrast 'She has a temperature of 38°C' is especially significant because the temperature has been measured to give a numerical reading. This is accepted as an objective measurement and would only be queried if there were reason to doubt the reliability of the thermometer or the competence of the nurse.

It is the uncritical use of mathematics and technical terms which is misleading and which can be dangerous – we can make bad mistakes if we use language we do not fully understand. But, if used with understanding, a technical language allows us to make concise and relatively precise descriptions in a form easily comprehended by other users of the language. We have to concede that scientific precision and conciseness must give us an incomplete account but, as we have seen (Chapters 2 and 4) *no* description can be complete. In one sense technical terms, which imply complex theories, are more likely to give a relatively comprehensive picture than are non-technical terms, dependent on less sophisticated theories; thus 'pH 3' gives more information than 'This is acid'. Scientific and technical descriptions do not necessarily *replace* those in ordinary language, they offer a sophisticated account that is highly significant and informative to those who understand the theories and therefore know the scientific language.

The use of technical terms and language is not confined to science; technical terms appear in cookery recipes, gardening books, accounts of games and of many hobbies. 'Roux', 'pruning', 'bishop's move' and 'penny black' are all technical terms as used in cooking, gardening, chess and stamp-collecting respectively. To understand them it is necessary to know something about these activities; the terms imply a background of theoretical as well as practical knowledge that Ryle calls 'quantity of luggage'; the 'luggage' carried by scientific terms is heavier, but the non-scientific luggage is like the scientific luggage in being part of theory:

> it is relatively easy for an ordinary Poker-player to explain in words the differences between the quantity and type of luggage carried by an expression like 'straight flush' and the quantity and type of luggage carried by an expression like 'Queen of Hearts'. But the corresponding task in some other fields is far from easy. Precisely how much more theoretical luggage is carried by such a term as 'light wave' than is carried by such a term as 'pink' or 'blue'? But at least one can discern very often that there is an important difference between one term and another, namely that one of them carries some of the luggage of a specific theory.[7]

As we have seen (Chapter 3), the theories involved in ordinary talk of physical things and words such as 'pink' and 'blue' are not generally formulated, we see *through* the theory to the fact; all our accounts of the world depend on theories but are also restricted by them. It is both a necessity and a limitation that a scientific or any technical language must give an account that is confined by its theory – its luggage. If we understand the theory the account is more precise and more comprehensible but inevitably we lose flexibility and we lose aspects not subsumed by the theory. Precision and comprehensibility conflict with flexibility and comprehensiveness. At one time, it was thought (certainly by many in seventeenth-century England) that the physical cosmos could be completely and accurately described in simple mathematical terms. God was a mathematician and it was the scientific ideal to discover the simple divine plan. Today we have come to accept that the inanimate physical world is complex, perhaps as complex as the world of human desires and intentions. Nevertheless though we have abandoned hope of finding that the world conforms to elegant mathematical laws we still seek to describe and explain with the minimum complexity. We need to impose order on what might be chaos and we are prepared to simplify in order to understand.

The sacrifice of comprehensiveness to comprehensibility is seen in the appeal to ideal models which feature in many scientific theories. Models such as the ideal gas, the frictionless fluid, the average man and the balanced economy are treated as entities though no observed phenomenon satisfies the model. It happens that the models of physical science approximate quite closely to what can be observed and that deviations from the model behaviour can be systematically allowed for. The models of the human sciences do not generally approximate so closely to particular observed cases and also the deviations cannot be so systematically accounted for. Therefore the model may be too remote from reality to have much practical use; this is to be discussed further in Chapters 8 and 11.

To summarise: well-defined terms, measurement and mathematical laws have been of very great value to physical science; one important criterion for distinguishing common-sense from scientific descriptions, that the latter are more precise, is demonstrated by the precision of mathematics and of defined technical terms. The same criterion should distinguish common-sense accounts from human-science descriptions but there is, to date, less scope for objective measurement and mathematical laws and the technical terms are more likely to be used ambiguously.

6

The D-N Model and the Concept of Law

In Chapter 1 it was stated that an important aspect of explanation consisted in relating the phenomenon to be explained (the *explicandum*) to some accepted regularity, that is to an empirical law. If this can be done we are on the way to producing a causal explanation as to *why* the phenomenon occurred and/or *why* it was as it was.

It should now be plain that all statements of regularity and law are supported by implicit or explicit theories. That water flows downhill, that bread nourishes, that chickens lay eggs are common-sense facts (albeit theory-laden) and they can also be regarded as laws. Likewise scientific facts: that gases expand when heated, that litmus turns pink in acids, that protein foods provide material for body cells, that Halley's comet will return every 76 years, all laden with scientific theories, can be regarded as laws. Thus empirical laws, like empirical facts, are theory-laden. Similarly the facts of human behaviour: that people seek food and water, that they laugh if tickled, that they seek revenge if wronged can be regarded as laws. But because the relevant theories are more tentative and the concepts more flexible we tend not to take human behaviour as being rigidly law-governed and our expectations are based on quasi-laws[1] rather than true laws.

A full causal[2] explanation, an explanation as to *why* things happen or *why* they are as they are, consists of a statement of an accepted law (or laws) and some other statement(s) relating the particular circumstances to it (or to them) so that from all these the *explicandum* can be deduced. The explanation is thus in the form of a logical argument from premises (the law(s) and particulars) to conclusion; this type of explanation is therefore called a Deductive-Nomological (D-N) explanation. An example is given by Popper:

> To give a *causal explanation* of an event means to deduce a statement which describes it, using as premises of the deduction one or more *universal laws*, together with certain singular statements, the initial conditions. For example, we can say that we have given a causal explanation of the breaking of a certain piece of thread if we have found that the thread has a tensile strength of 1 lb and that a weight of 2 lbs was put on it. If we analyse this causal explanation we shall find several constituent parts. On the one hand there is the hypothesis; 'Whenever a thread is loaded with a weight exceeding that which characterizes the tensile strength of the thread, then it will break'; a statement which has the character of a universal law of nature. On the other hand we have singular statements (in this case two) which apply only to the specific event in question: 'The weight characteristic for this thread is 1 lb' and 'The weight on this thread is 2 lbs'. . . .
>
> It is from the universal statements in conjunction with initial conditions that we *deduce* the singular statement, 'This thread will break'. We call this statement a specific or singular *prediction*.[3]

The general rules for an acceptable D-N explanation were formulated by Hempel[4] and Oppenheim and have been recapitulated by Ryan:

> a successful explanation has to obey three requirements. The first is the formal requirement that the statements laying down the laws and initial conditions should entail the statement laying down the conclusion; the second is the material requirement that the premises should be true – or more cautiously that they should be well corroborated; the last is a consequence of these requirements, that the *explanans* should be empirically testable, by being open to refutation should it predict what is not the case.[5]

It is to be noted that Popper takes prediction to be the essence of causal explanation and we shall see that Ryan also regards explanation and prediction as complementing each other in the D-N model. This view will be criticised later but first we need to consider the concept of law itself since the strength of any D-N explanation is directly related to the reliability of the law or laws on which it rests. Laws supporting empirical explanations must be empirical laws and, as Ryan says, they must be open to corroboration (and, of course, possible refutation) through appeal to observation. In this respect they differ from mathematical and logical laws and also from non-empirical definitions such as 'All lodestones attract iron'. 'All lodestones attract iron' is no more an empirical law than is 'All bachelors are unmarried'. Neither statement is confirmed by *experience* and experience (observation) cannot show either statement to be false. We may grant that, originally, the fact that lodestones attracted iron was an empirical discovery but that discovered property has now become part of the collection of properties that *define* lodestones. Hence, just as if we found that a man whom we had been told was a bachelor was married, we should say that he was not, after all, a bachelor – we should *not* say that the statement 'All bachelors are married' was false – so, if we found that what we had been told was a lodestone did not attract iron we should say that it was not, after all, a lodestone – we should *not* say that the statement 'All lodestones attract iron' was false.

'Attracting iron' is now accepted as a defining property of lodestones, but there are many examples of properties which are of ambiguous status, that is to say it is not clear whether they are empirical properties (which observation might show not to hold universally such as, for example 'white' as a property of swans) or whether, like 'attracting iron' as describing lodestones, they are defining properties. Hence it is not clear whether the corresponding statements are empirical laws, subject to refutation or qualification, or are irrefutable definitions. For example it is not clear whether 'Copper melts at 1083°C' and 'Hydrogen is an explosive gas' are empirical laws or definitions. They have equivocal status because we have such absolute reliance on our theories of uniform association of properties (see Chapter 2) that we do not envisage that *as empirical propositions* they could be false. Therefore they come to be *used* as definitions as well as being regarded as statements of empirical fact. We do not seriously entertain the suggestion that the laws will be refuted but what grounds do we have for such confidence?

It was in the eighteenth century that David Hume pointed out that we have no *logical* reason to expect the observed association of properties of objects and materials and the observed sequences of events to continue and that therefore we have no *logical* certainty that even the best-corroborated laws, those that have shown no exceptions up to the present time, will hold in the future. Hume suggested that our trust in the uniformity of nature, our confidence that the future would resemble the past was the psychological effect of experience – custom and habit – but it could not be justified by reason. There have been many attempts to counter Hume's point and to solve what has come to be known as 'Hume's problem of induction' but none of them has been successful.[6] However, we can offer reasons for setting Hume's scepticism to one side when we seek causal explanations by appeal to law, for though we cannot say that it is logically impossible for the regular order to end we can *rule against* disruption when we set out to seek to describe and explain what happens in the world. Any search for empirical knowledge must rest on the assumption that there *is* order and that it is possible to learn about that order by using past experience. The fundamental importance of metaphysical principles was referred to in Chapter 1 (p. 2); *some* have to be accepted if we are to have any concepts of phenomena; without assuming regular association of properties we should not be able, as Winch says,[7] to describe our world. *A fortiori* we should not be able to explain it.

This does not entail that we must regard all our current beliefs in the truth of common-sense and scientific generalisations as sacrosanct. What it does entail is that any corrections made will *not* be made because we think there is a change in the order of nature, they will be made because we discover that our observations are faulty or inaccurate, or that we had not appreciated the effect or influence of certain other phenomena, or that we have decided to organise our data and interpret our observations in a new way. The changes are due to *us*, not to the world. Thus we accept that our D-N causal explanations rest on laws that are subject to correction in the light of future experience and future interpretations, but we also assume our laws depend on a regularity that we rely on as a matter of metaphysical principle and which is not subject to change.

However, though we can assume regularity and therefore can ignore Hume's problem we have to deal with another problem, that of distinguishing genuine empirical *laws*, laws of nature, from empirical generalisations that just happen to be true. It was Descartes who first introduced the term 'law of nature'; he and most of his contemporaries held that God had laid down laws which determined the behaviour and the interactions between all physical objects and materials and the course of all physical events. By contrast accidentally true generalisations simply described the way things happened to be.[8] Hence laws of nature differed from accidentally true generalisations because they had a prescriptive force; they were analogous to human laws but with the important distinction that the divine laws of nature *could not* be broken, all the material universe was subject to them. In the words of the hymn:

> Praise the Lord! ye heavens adore Him,
> Praise Him, Angels in the height;
> Sun and moon rejoice before Him,
> Praise Him, all ye stars and light.

> Praise the Lord! for he hath spoken;
> Worlds His mighty voice obeyed:
> Laws, which never shall be broken,
> For their guidance He hath made.

> Praise the God of our salvation;
> Hosts on high His power proclaim,
> Heaven and earth and all creation,
> Laud and magnify His name![9]

Today we do not take scientific laws (laws of nature) to be divine decrees that we have discovered. Neither do we regard them as *rules* guiding the course of events; they are held to be our *descriptions* of the world. But there remains an element of prescriptive force for they are *not* thought to be just descriptions of the way things happen to be, they tell us not only what is but also what was and what will be. As Ayer says:

> Once we are rid of the confusion between logical and factual relations, what seems the obvious course is to hold that a proposition expresses a law of nature when it states what invariably happens. Thus, to say that unsupported bodies fall, assuming this to be a law of nature, is to say that there is not, never has been, and never will be a body that being unsupported does not fall. The 'necessity' of a law consists, on this view, simply in the fact that there are no exceptions to it.[10]

Ayer points out that statistical laws can be regarded in the same way as universal laws: 'All men are mortal' and '75% of hybrid crosses show the dominant trait' have the same force and the same 'necessity'.[11]

Another characteristic of laws, as opposed to accidentally true generalisations, is that they can be projected to cover hypothetical events, that is events that might have occurred but did not in fact occur. Laws can predict what would happen and can tell us what would have happened and so, unlike accidentally true generalisations, they support unfulfilled conditionals. In another work Ayer gives an example:

> let us suppose that a meeting is held at which some motion is passed unanimously with no abstentions. Then it will be a true generalization of fact that all the persons present at the meeting voted in favour of the motion. Given the characters and opinions of the persons in question this may not be wholly accidental, but the generalization still falls short of being a generalization of law. We can infer with regard to anyone you please that if he was present at the meeting, he voted in favour of the motion, but we cannot infer with regard to anyone you please that if he had been present he would have voted in favour of the motion, for among those who were not present there might very well be some whose character and opinions were such that if they had been present, they would have voted against it. On the other hand, if we consider the generalization that all those present at the meeting were warm-blooded, we think we can safely infer with regard to anyone you please not only that if he was present he was warm-blooded, but that if he had been present he would have been warm-blooded. It being a generalization of law that all men are warm-blooded, we can extend it to merely possible cases in a way that we cannot extend a generalization of fact.[12]

There are important and interesting problems arising from the need to distinguish between genuine laws of nature and accidentally true generalisations of fact. If the law can itself be deduced from a higher level theory its status as a scientific law, as a genuine law of nature, is made more secure. But there are some generalisations that we regard as laws of nature even though they are not supported by a higher-level theory, for instance 'All men are mortal'. We can 'explain' a man's death by setting up a D-N argument with the premisses:

All men are mortal
This was a man
Therefore he had to die

but such an explanation would, in general, be thought to be very unsatisfactory. That same D-N argument can also support a very reliable prediction:

All men are mortal
This is a man
He will die

The prediction is barely worth making because it is not doubted; only death and taxes are certain. However, the D-N example serves to illustrate what was referred to at the beginning of this chapter, namely that many philosophers regard prediction and explanation as complementary aspects of a D-N argument. Thus Ryan says:

> the only differences between explanation and prediction lie in whereabouts in the deductive argument we begin. With the case of explanation, we begin with the event to be explained, i.e. with the conclusion of the argument, and look for the appropriate generalizations and initial conditions from which to deduce this conclusion. In the case of prediction, we begin with the generalization and the initial conditions and then go on to forecast the coming occurrence.[13]

For some philosophers the outstanding feature of explanations is the fact that they have predictive force. Hempel says:

> It is this potential predictive force which gives scientific explanation its importance: only to the extent that we are able to explain empirical facts can we attain the major objective of scientific research, namely not merely to record the phenomena of our experience, but to learn from them, by basing upon them theoretical generalizations which enable us to anticipate new occurrences and to control, at least to some extent, the changes in our environment.[14]

Another writer, Rescher, while acknowledging that there are other aspects of explanation, also stresses the importance of prediction:

> While prediction and control cannot be taken as exclusively constituting the 'aims of science', they do represent criterial factors by reference to which the superiority of the scientific framework of explanation can be established.[15]

Clearly prediction and predictive power can be important to us, but does our search for understanding and knowledge of the world reduce merely to a search for correct causal explanations and reliable predictions? Is this even *primarily* what we seek? Undoubtedly the ability to predict and control can be of major concern in some situations[16] but if we want understanding a D-N explanation seems inadequate, sometimes very inadequate as the example of mortality shows. In addition the relation between prediction and explanation is not symmetrical. As we saw, a prediction so reliable that it is hardly worth making can have very weak explanatory power.

The inadequacy of the D-N model is stressed by Scheffler. He uses the term 'causal explanation' in the same sense as does Popper; he points out that a law such as 'All samples of copper will conduct electricity' allows us to predict that a given sample of copper will conduct electricity. This is analogous to predicting that a given man will die from the law 'All men are mortal'.

> We may well predict that the sample of copper before us will conduct electricity under test, on the ground that all previous samples have, but we do not *explain why* it does so by listing all those previous copper samples which have, in fact, conducted electricity.[17]

This example is not quite fair because, as has already been pointed out, the law used in a D-N explanation may itself be supported by an explanatory theory. It *is*, I think, a fair criticism of the model as given in Popper's example (p. 53) for Popper's law is indeed 'bare'; just 'Whenever a thread is loaded with a weight exceeding that which characterizes the tensile strength of the thread, then it breaks'. There is no suggestion of an explanatory theory of that law. But if a law used in a D-N explanation is related (or can be related) to an explanatory theory, that theory will 'fill out' and enrich both the law and the explanation and so make the latter more satisfying. For example if we know that metals that are conductors of electricity do so because electrons can pass freely between their atoms the Scheffler example is less 'bare' for 'Copper is a good conductor of electricity' is or can be understood as a minor premiss of a fuller D-N explanation.

However, Scheffler's criticism does show that a correct D-N explanation can be inadequate, for it need not introduce any widening of outlook or of concepts.[18] The theory that an electric current is a flow of electrons goes well beyond what is required for the logical deduction that *this* piece of copper is a conductor; all we *need* is the law 'All samples of copper conduct electricity'. Hence though a bare causal explanation can support a reliable, indeed a firm, prediction it does not necessarily give us all we expect. This is just what Scheffler objects to about the D-N model. He considers that the conceptual aspect of explanation is the most important aspect and he argues that the search for physical causes serving as a basis for prediction is never a primary concern in theoretical (as opposed to applied) science. He takes a very different view, therefore, from Hempel about scientific explanation. He argues that those working in the fields of physical *and* human sciences seek conceptual understanding, an understanding that gives a (metaphorical) picture in which the *explicandum* is 'placed'. For him descriptive and conceptual aspects of explanation coalesce in a broader scheme:

Whatever particular account is given of the reason, it seems true that in science, as distinct from practical affairs, neither the search for causes, nor causal explanation, is *primary* in any likely sense that may be assigned to this vague term. Unlike our concern with control in practical affairs, our concern in science is not directed particularly towards the future, but frequently involves employment of general principles in the effort to substantiate past events on the basis of later events, as, for example, in history, cosmology, archeology, paleontology and geology. Non-explanatory [Scheffler means 'non-causal'] substantiation may be illustrated by the use of various methods of dating historical remains, or the analysis of rocks in connection with principles of radioactivity to determine the age of the earth. . . . The positing of past events is illustrated by the use of all available current information to paint a picture of earlier times, say the life of medieval people; the historian is often concerned just with painting such pictures, rather than with [causal] explanation proper, and the same holds true of investigators in other realms.[19]

It may seem that Scheffler thinks that the search for causal explanation and predictive power is completely separable from the search for conceptual understanding. I suggest that this is not so: not only need they not be separate, they rarely are separate. Indeed except for historical explanations of some kinds it would be very difficult, perhaps impossible, to separate them. I prefer to interpret Scheffler's argument as being directed against the Hempel view that explanation consists *essentially* of a search for causes and for predictive power; I suggest that Scheffler regards the D-N model as being inadequate rather than inappropriate.

The inadequacy of the D-N model is particularly marked in the human sciences. This is not only because, as in the physical sciences, it may not increase conceptual understanding[20] but also because it is unlikely to be so applicable even within its limitations. As was mentioned at the beginning of the chapter, the laws relating events in the fields of social science and in human behaviour tend to be less reliable, they are better described as 'quasi-laws', and a D-N argument incorporating quasi-laws cannot provide so conclusive a conclusion (and therefore so reliable an explanation) as an argument incorporating laws of nature proper. In addition it is generally much more difficult to specify the relevant initial conditions affecting social and individual human situations so that the minor premises of an explanatory D-N argument are also less sound. Some philosophers maintain that explanations in the social sciences cannot be analysed in the way that they are analysed in the physical sciences and that the D-N model is not applicable. We shall discuss this further in Chapter 11.

But first we should discuss the general aspects of explanation and then we can return to reassess the part played by *causal* accounts in relation to the descriptive role of explanation and in relation to the search for understanding.

7

Aspects of Explanation: How, Why and Wherefore

We have seen that explanations conforming to the D-N model provide causal accounts as to why certain events occur, and that also they are a source of predictions. In addition to their causal/predictive function explanations have two others: a descriptive function, answering the question 'How?', and a conceptual function, answering the question 'Wherefore?' An explanation may be made primarily to fulfil just one function but most will embody material to fulfil at least two functions and some will fulfil all three. The different functions of an explanation are its different aspects and cannot be separated within the explanation, but they can be discussed separately.

The descriptive aspect might not be thought to be explanatory for just to present a description is not normally to provide an explanation. Yet as we shall see, descriptions can be explanations and indeed we may come to support the view that, in an ultimate sense, all explanations must be treated as descriptions – this will be discussed in our final chapter. However, even in a non-ultimate sense, description is an important aspect of explanation for we must be able to describe phenomena and events before we can suggest causes and before we can relate them to other phenomena and events. In addition the *way* we describe, and this depends on what caught our attention as well as how we interpret the observation, will partly determine the causal and conceptual aspects of the explanation. Thus description is a first and essential basis of any explanation as well as perhaps being the culminating end of explanation.

Our discussion of the causal/predictive aspect of explanation in the previous chapter showed that a satisfying causal explanation does more than rest on a simple law of uniformity; the law (or laws) will be supported by explanatory theories and so can lead us to a deeper understanding by 'placing' the *explicandum* in a broader pattern of events answering the conceptual question 'Wherefore?' as well as the causal question 'Why?' Satisfaction of the conceptual search must involve implicit or explicit appeal to higher-level theories. An example of explanation in all its aspects is given by Salmon:

> A colleague, to whom I shall refer . . . as the 'friendly physicist', . . . noticed a young boy sitting across the aisle [of the aeroplane] holding on to a string to which was attached a helium-filled balloon. He endeavoured to pique the child's curiosity. 'If you keep holding the string just as you are now,' he asked, 'What do you think the balloon will do when the airplane accelerates before take off?'. The question obviously had not crossed the youngster's mind before that moment, but after giving it a little thought, he expressed the opinion that the balloon would move toward the back of the cabin. 'I don't think so', said the friendly physicist, 'I think it will move forward.' The child was eager to see what would happen when the plane began to move. Several adults in the vicinity were, however, skeptical about the physicist's prediction; in fact a stewardess offered to wager a miniature bottle of Scotch that he was mistaken. The friendly physicist was not unwilling and the bet was made. In due course, the airplane began to accelerate, and the balloon moved toward the front of the cabin.[1]

The immediate question was a request for a prediction as to what would happen to the balloon when the airplane accelerated forward, 'How would the balloon change?' The boy's (wrong) answer was of the right type for it was related to the *horizontal movement* of the balloon; it was the kind of answer that was expected. Other kinds of change: of density, size, colour, shape or vertical movement were not anticipated. *Everyone*, including the friendly physicist, was (directly or indirectly) already conditioned by current physical theory to expect horizontal movement and nothing more. Hence though the boy and the passengers (save for the friendly physicist) were surprised when the balloon moved forward rather than backward it was a limited surprise.

We have seen that currently accepted theories limit as well as guide what we observe[2] and what we describe. We tend to ignore small changes that we do not anticipate – no one might have noticed if the balloon had become slightly larger or smaller for instance. In this respect the friendly physicist might have expectations as limited, perhaps even more limited, than the boy or the other passengers. Nevertheless, since we cannot observe and take note of everything we have to select (consciously and unconsciously) from an indefinitely large number of possible accounts and we have to accept the not inconsiderable risk of being misled by the theories that guide our selection and of overlooking a change that is not part of those theories. The boy's prediction was shown to be false and he had thereby gained more knowledge but, at least at first, he and the other passengers would have been more puzzled than they had been in their original comfortable state of ignorance. In Chapter 1 it was stated that familiar and unexpected events stimulate curiosity much more than the familiar and expected; no one save the friendly physicist would have been interested if the balloon had moved backwards as the child had predicted.

Now a D-N causal explanation based on the law 'Helium balloons in an aeroplane move forward when the aeroplane accelerates forward' provides a causal explanation (analogous to that given by Scheffler, see Chapter 6, p. 59). But *that* explanation would not satisfy curiosity even though (or perhaps even if) the law was not doubted and was therefore accepted as a reliable basis for prediction. To provide a satisfactory explanation the question 'Wherefore?' as well as the question 'Why?' must be answered. Only the friendly physicist had been aware of the important difference the low density of helium (the balloon on the string was upright in the air) would make. He knew that inertia would move the balloon in the opposite direction to most objects, since most objects are more dense than air. Just as objects less dense than air behave differently when not supported – they rise, whereas objects more dense than air fall, so an object less dense than air behaves differently in a forward-accelerating air-filled container and moves forward whereas the objects more dense than air move backwards. A D-N explanation implying or incorporating the fuller theory of inertial motion provides a causal and a conceptual explanation thereby giving increased understanding of inertia and inertial motion. This deeper explanation also gives more possibility of control; in the example given it is shown that in order to stop the inertial motion (in air) of *any* object less dense than air, restraint must be to potential movement in the *same* direction as the accelerating container.

In many situations we wish to be able to predict in order to control events, and the fuller the causal explanation, that is the more of the conceptual it incorporates, the more likely is the possibility of that control. For example we can predict increase and loss of human body weight knowing that weight is causally connected to diet but theories suggesting that it is also affected by exercise and metabolic rate, by percentage of brown fat and by psychological factors give us greater understanding and also make control more possible. Likewise we can predict and, hopefully, control the spread of diseases by using our knowledge of simple cause and effect and further theories explaining the more simple regularities. Unfortunately there are many events that we cannot control even though we know something of their causes and can predict them, for example the coming of a drought, or of a hurricane or an earthquake. We are able to make very accurate predictions of astronomical events such as eclipses, the return of comets, and the positions of the stars, though we have no control over them. Even in circumstances where there is no control the power to predict can be useful because forewarning may give time to reduce danger and damage, to take precautions and/or to take people from a danger area. What we want from an explanation will depend on the situation; as Salmon says:

> Theoretical science furnishes both explanations and predictions. Some of these predictions have practical consequences and others do not. When, for example, scientists assembled the first man-made atomic pile under the West Stands at the University of Chicago, they had to make a prediction as to whether the nuclear chain-reaction they initiated could be controlled, or whether it would spread to surrounding materials and engulf the entire city – and perhaps the whole earth – in a nuclear holocaust. Their predictions had both theoretical and practical interest. Contemporary cosmologists, for another example, would like to *explain* certain features of our universe in terms of its origin in a 'big bang'; many of them are trying to *predict* whether it will end in a 'big crunch'. In this case the predictive question seems motivated by pure intellectual curiosity, quite unattached to concerns regarding practical decision-making. Whether a helium-filled balloon will

move forward in the cabin of an airplane when the airplane accelerates, whether a nuclear chain reaction – once initiated – will run out of control, and whether the universe will eventually return to a state of high density are all matters of legitimate scientific concern.[3]

This quotation highlights the importance of the 'wherefore' aspect of explanation: *'whether* a nuclear reactor will become uncontrollable, *whether* the universe will run down' are questions that seem to require more than an answer that is a bare prediction. The first question implies concern with decisions about the making of nuclear reactors, the second question arises not only from intellectual curiosity but from thoughts about our significance (or insignificance) in relation to the cosmos. Von Wright[4] refers to *Geisteswissenschaften*, a term which for him embraces explanation and understanding (for us, both are part of the 'wherefore' aspects of explanation, see Chapter 1, p. 2). Von Wright says that *Geisteswissenschaften* was introduced by Dilthey in the late nineteenth century, and was the German translation of 'moral science' as used in Mill's *Logic*.[5] Hence the 'wherefore' aspect of explanation does carry a penumbra of ethics and both questions and answers will embody moral values in a more direct way than do questions and answers relating to description or cause ('how' and 'why' matters). It is not surprising that the 'wherefore' aspect of explanation is especially prominent in explanations of the actions of agents[6] for agents are almost invariably human and are morally aware. According to von Wright[7] the hermeneutic-dialectic philosophy and methodology is particularly concerned with explanation in terms of intention. Hermeneutics means the art of interpretation – originally the interpretation of documents, but the hermeneutic philosophy developed in the 1960s is involved with the interpretation of people's behaviour. It has as its field language-orientated notions and the significance of language and meaning and it is based on thought and discussion rather than experiment and observation. It is therefore quite different from traditional scientific inquiry, which seeks causal laws and which the D-N types of explanation discussed in the previous chapter. Von Wright says: 'hermeneutic philosophy defends the *sui generis* character of the interpretative and understanding methods of the *Geisteswissenschaften*'.[8]

The hermeneutic philosophy is more concerned with semantics

than with empathy and the operation of *Verstehen* but it does seek to provide part of the answer to 'wherefore' questions and the 'wherefore' aspect of explanation is as much to be considered as the descriptive ('how') and the causal ('why'); it can make a significant contribution. However, causal explanations are of major importance, and we must now return to further consideration of the role of laws.

8

Laws and Quasi-laws

The confidence we have in a D-N explanation and in the related prediction is clearly directly related to the reliability of the law (or laws) on which it depends. Not that reliability alone is sufficient for, as we have seen, if a law asserts a regularity that is not itself explained by a higher level theory the D-N explanation is unlikely to be satisfying. The blackness of a raven may be explained by the D-N argument:

> All ravens are black
> This is a raven
> Therefore it is black

but this would not generally be thought to be satisfactory. It is comparable to the explanation of mortality (see Chapter 6, p. 58). Curiosity about the colour of *a* raven is, in general, curiosity about the colour of *ravens* so that appeal to the law 'All ravens are black' is not taken to be a true explanation. Our present knowledge suggests that there is likely to be a 'black-producing' gene and we expect an explanation of the colour of ravens to refer to that gene (or perhaps genes) and its (or their) characteristics. The explanation of the colour of these birds becomes part of a wider pattern of explanation of colour and of other inherited qualities of ravens and of all living things. Furthermore we might be inclined to question the simple law 'All ravens are black' for perhaps there have been albino ravens. An explanation involving a higher-level genetic theory is likely not only to explain the lower-level law but also to explain the exceptions. As a rule empirical generalisations unsupported by higher-level theory are not thought to be as *reliable* (let alone as satisfying) as generalisations that are explained (and therefore supported by) such a theory. 'All men are mortal' is unique; the exception that proves the rule. Elsewhere[1] I have suggested that empirical generalisations should only be considered to be laws when they are supported by and can be derived from a higher-level theory. In this text we use the term 'law' or 'law of nature' to signify a well-attested generalisation that has held, holds and will hold and that also supports (may be applied to) hypothetical events.[2] On these criteria 'All men are mortal' and 'All ravens are black' are laws, laws of nature; but there are a relatively small number of simple generalisations with this status. In general they do not possess the force of factual necessity that laws explained by and derived from theory have.

We seek to explain regularities observed, not only universal generalisations such as 'All objects denser than air fall to the ground' but also statistical generalisations such as '25% of heavy smokers die of lung cancer before the age of 60'. The universal generalisation, a non-mathematical simplification of Galileo's law, is explained and may be derived from Newton's theory of gravitational force. Before this explanation was available many philosophers, Descartes for example, had treated what we now call Galileo's laws of falling bodies as mere generalisations. They did not have the status of laws even though they were invariably confirmed by observation. The statistical generalisation '25% of heavy smokers die of lung cancer before the age of 60' has acquired a firmer law-like status not just because there are more observations to confirm it (and none that tend to refute it[3]) but also because we have evidence to support theories that tobacco tar has a carcinogenic effect. Likewise Mendel's law of hybrid crossing '75% of crosses show the dominant trait' has had law status since it became supported by theories[4] of the transmission of characteristics by genes.

In the fields of psychology and the social sciences the D-N explanation is often used even though, as we saw in Chapter 6, it is likely to give a less complete or satisfying explanation than it would give in the fields of physical science. It is still important to be able to appeal to established regularities, that is to laws. We need to consider what kinds of laws are used and to assess them in relation to the laws that feature in explanations pertaining to the physical sciences.

Carnap gives an example of a D-N explanation of human behaviour and shows how it depends on an appeal to laws:

> We ask little Tommy why he is crying and he answers . . . 'Johnny hit me on the nose'. Why do we consider this a sufficient explanation? Because we know that a blow on the nose causes pain and that, when children feel pain they cry. These are general physiological laws. They are so well known that they are assumed even by Tommy when he tells us why he is crying.[5]

We can set out the two laws:

(1) A blow on the nose causes pain.
(2) Children cry when they are in pain.

They are analogous to the laws of physical science in that they can be explained by and can be derived from high-level theories but they do not have the same straightforward applicability as physical laws. They must almost always be qualified to allow for the nuances and complexities of situations where they are or could be applicable. Nevertheless they are to be distinguished from accidentally true generalisations for we can allow that they are, with appropriate qualifications, universally valid now, were so in the past and will be in the future; we can also allow that they are projectible to hypothetical events. They differ from laws of the physical sciences in that there are no overall rules as to how they may need to be qualified in different circumstances. For example the particular conditions, what were called the 'initial conditions' will directly affect the qualifications required so the laws are not independent of those conditions; we may say that quasi-laws indicate tendencies rather than asserting rigid regularities. 'A blow on the nose causes pain' is a quasi-law because there are circumstances in which it would not apply: the attention of the individual might be directed elsewhere, he or she might be too angry or excited to feel the pain, or might have a high pain threshold. Likewise 'Children cry when they are in pain' is a quasi-law: some children are more stoical than others, the age of the child is relevant, whether she is seeking attention, or trying to show that the blow had no effect, and so on. None of these factors can be precisely measured or assessed yet, since they qualify the quasi-law that law can only indicate tendency. Rescher gives another example:

the quasi-laws that in a U.S. off-year election the opposition party is likely to gain. This is certainly not a general law, nor is it a mere summary of observed statistics. It has implicit qualifications of the 'other things being equal' type, but it does claim to characterize the course of events 'as a rule' and it generates an expectation of the explainability of deviations. On this basis, a historical (or political) law of this kind can provide a valid though limited foundation for sound predictions.[6]

Many causal explanations in the human sciences are based on quasi-laws of this type. The force of the quasi-law is not only affected by the complexity of the particular circumstances (the initial conditions) but it can also be affected in a general way by circumstances external to the particular situation, circumstances that can seriously modify and that may come to undermine the quasi-law. A necessary characteristic of laws and quasi-laws is that they will hold in the future (see p. 56) but this implies that the relevant external circumstances will be unchanged. The laws of human behaviour, as opposed to the laws of physical science, very often depend on external conditions that are known to be constantly changing and so they do not have the same permanence, the same independence of time. For example, most of us believe that the physical law 'The orbits of the planets are ellipses' will hold for centuries in the future as it has in the past though we may also know that should several large comets enter the solar system the planetary orbits could change. We do not regard the physical law as impermanent because we think it highly unlikely that the external circumstances will change. By contrast changes in external circumstances that could undermine a quasi-law of social science are very likely. For example the quasi-law 'Trade Unions support the Labour Party' has not the same force as when the Labour Party was founded. The support of the unions as highly dependent on the views of a relatively small number of key individuals and their decisions are affected by external political events. In the latter part of the twentieth century the rise of the SDP, the formation of the Alliance Party, changes in both Tory and Labour Party policy and events in other countries, the USSR, South Africa, the USA and France, are all external factors that may have a profound influence on the individuals whose decisions will show whether the quasi-law operates.

Although social scientists must rest content with quasi-laws it does not follow that there can be no reasonably reliable predictions and that no sensible or significant general statements can be made. Discussion can be helpful and progress can be made as long as it is appreciated that circumstances are complex. A further difficulty, referred to in Chapters 2 and 5, is that the words and terms used in quasi-laws may not have precise definitions or stand for clear concepts and therefore the language of the human- and especially social-science theories and explanations is context-dependent and less objective than the language of physical science. It does not follow that human-science explanations cannot be as satisfying as those of physical science; the latter aspire to offer objective and complete accounts, the former suggest complementary pictures dependent on complementary theories.

> Language and concepts mirror the world in a direct way, . . . But . . . how we conceptualise the world depends on our theoretical interests. Take the concept of the firm. On one view, the firms are profit-maximising, the concept of the firm covers most multi-national corporations and corner shops, but not some nationalised industries. . . . On another view, the firm in an advanced capitalist society is a giant, highly integrated and technologically advanced economic institution, the concept of the firm covers most multi-national corporations, ... Are multi-national corporations more like nationalised industries than corner shops? Which concept of the firm is more useful for understanding society? These questions cannot be approached through naive gazes at the social world. They are meaningless unless we consider the different theories of the firm sustained by the different concepts.[7]

The complexity of particular circumstances, the complexity of variation in external factors, the fluidity of human affairs and the confusions that may arise from ambiguous terminology and latent differences in concepts and theories can lead an enthusiast and also a cynic to formulate laws that can (*post hoc*) explain any situation or event that occurs. Then any regularity appealed to is not even a quasi-law, it is spurious and any explanation based on it is spurious. In the next chapter we shall consider the nature of spurious explanations.

9

Spurious Explanations

Spurious explanations are those that appear to give greater understanding by suggesting causes (and apparently supplying bases for predictions), and/or by suggesting theories that seem to set the *explicandum* in a larger conceptual frame. In Chapter 5 (p. 46) it was noted that concepts that are vague may make explanations so loose that they can be regarded as compatible and as incompatible with the same situation. 'Vandalism is caused by unemployment' can be interpreted in such a way that advocates of the explanation can always find evidence to support it and objectors may use the same evidence to refute it. These explanations are not being criticised because they are thought to be *wrong* but because they are thought to be spurious. Critics say that, for example, the Freudian theory that neuroses are caused by an inferiority complex is spurious since it turns out that *any* observed behaviour, submissive or aggressive, can be 'explained' by appeal to this complex. Since the theory can 'explain' any behaviour that is observed it cannot be used to predict what will be observed; the 'explanation' is always *post hoc*. An essential feature of a genuine empirical explanation is that it arises from theories (and laws) that *exclude* certain outcomes[1] and so allow us to predict (within prescribed limits) what will happen. Explanations in the human sciences are particularly liable to be spurious, or to verge on the spurious, because, as we have seen, the concepts and the facts tend to be more flexible, indeed some flexibility is desirable. Nevertheless they cannot be so vague that they can accommodate all outcomes – some limits must be set.

We shall return to this later, but let us begin by considering explanatory theories that are regarded as spurious by most scientists. Let us consider explanations based on astrological theories that, from the movements and positions of the heavenly bodies, the actions, characters and even future behaviour of human beings can be explained and, at least to some extent, predicted. Though serious astrologers would not take the predictions appearing in certain daily papers and magazines as true astrology it is clear that many readers give such crude theories some credence, for if they did not attract purchasers these 'predictions' would not be published.

Astrology can be regarded as a pseudo-science, a basis of pseudo-explanations, not only because there are no university departments[2] in astrology, but also because it has, when objectively assessed, been of no direct help in providing explanations and making predictions. This does not mean that it has been of no use at all: in so far as it has stimulated observation of the heavenly bodies, it has helped astronomy; but this is an indirect effect.[3] Moreover, relatively recently there has been some serious investigation of the claims of the more genuine astrologers (not those writing for the popular press). These astrologers make predictions that are not vague and ambiguous and therefore their predictions (and so their theories) can be tested. It may still turn out to be the case that they are *false*, and therefore do not provide an acceptable explanation, but at least they are not a source of spurious explanations.[4]

Genuine explanations may be true or false; spurious explanations are so framed that they cannot ever be shown to be false: that is the predictions derived from them, where we can compare what is predicted with what is observed, can *always* be interpreted to account for what happens *post hoc*. Let us consider the following examples taken from the *Daily Mirror* of 25 July 1986:

IF TODAY IS YOUR BIRTHDAY: A year of great changes lies ahead. You will meet new friends and develop fresh interests. Work opportunities will be plentiful.

★ **ARIES (March 21–April 20):** Negotiations for a job are concluded successfully. Home life looks like being busy, with many wanting your time. But you'll also enjoy a peaceful break to enable you to pursue your own interests.

☆ **TAURUS** (April 21–May 20): **You'll be busy with letters and phone calls in a bid to catch up on news. But leave time for fun with friends. Don't be afraid to make quick decisions. Most of them will be right and profitable.**

★ GEMINI (May 21–June 20): Whether you're hard at work or concentrating on your social life, you'll still be lucky with money. Some will receive confirmation of a pay rise, though they will have to take on extra responsibilities. If you're asked to organise an event in the near future, accept.

☆ CANCER (June 21–July 20): **It won't matter what trials you face, you'll find it easy to be tolerant and adaptable. This will please your partner's relatives. Plans for a visit to a new place will go well. You hear good reports of an old friend.**

★ LEO (July 21–August 21): This will be a good day for sorting out investment problems. There may be news of a windfall – though only a small one – to cheer you. You'll be able to handle someone's worries in a reassuring and subtle fashion.

☆ VIRGO (Aug. 22–Sep. 22): **It's one of those days when even a traffic warden might take pity on you and tear up a parking ticket. You may be thinking seriously about a marriage proposal but you shouldn't rush your answer. Friends prove their worth.**

★ LIBRA (Sep. 23–Oct. 22): A cheerful approach to life – particularly the areas that can bring you high rewards – is your greatest ally. You'll be able to work quickly and let others know exactly what you have in mind.

☆ SCORPIO (Oct. 23–Nov. 22): **A carefully planned trip could bring you closer to someone you've been pursuing. Don't hold back. Go as far as a candlelit dinner if funds will allow. A flamboyant approach will succeed more effectively than subtle hints.**

★ SAGITTARIUS (Nov. 23–Dec. 20): Though you may want to take a trip, the family and home will demand your attention today. Parents or older relatives will help you and you will assist them. Those hoping to move should be pleased by today's news.

☆ CAPRICORN (Dec. 21–Jan. 19): **Cut short conversations with those who do not think as quickly as you. Your impatience will show. You need to bounce ideas around with those who understand you best. Watch out for a minor financial hiccup.**

★ AQUARIUS (Jan. 20–Feb. 18): There's a danger that you can take practicality and efficiency too far. Enjoy the luck that surrounds dealings involving money. But leave some time – probably this evening – for enjoying yourself.

☆ PISCES (Feb. 19–March 20): **Make an effort to brighten up the day for yourself and those closest to you. Try to avoid being cooped up in the company of just one or two people. Plans made now will add an extra sparkle to tomorrow.**

There is much sound general advice here and the optimistic tone will undoubtedly cheer many readers and is almost certainly intended to do this: there is no hint of anything unpleasant occurring to anyone and even minor setbacks are presented as if avoidable. Moreover the 'forecast' for each zodiac sign could relate to anyone, especially as each is carefully qualified; for example *'looks like* being busy' (Aries); *'Some* will receive confirmation of a pay rise' (Gemini); 'There *may be* news of a windfall' (Leo); 'A carefully planned trip *could bring you* closer to someone you've been pursuing' (Scorpio); 'There's a *danger that* you can take practicality and efficiency too far'. There are no firm predictions here and it is hard to think of anyone who buys a newspaper in England, the *Daily Mirror* or any other, to whom these predictions could not apply. Thus astrology, at least as seen here, can be nothing but a source of pseudo-explanations and predictions lacking in content.

Now we will consider a less crude example, psycho-history. In contrast to newspaper astrology, this subject does have academic supporters. The following account is based on, and the quotations come from, an article written in *Encounter* (March 1979), by Professor Michael Shepherd, Professor of Epidemiological Psychiatry at the Institute of Psychiatry, University of London. One advocate of psycho-history is Professor William Langer, who lectured on the subject to the American Historical Association in 1957:

the newest history will be more intensive and probably less extensive. I refer more specifically to the urgently needed deepening of our historical understanding through exploitation of the concepts and findings of modern psychology. And by this, may I add, *I do not refer* [Shepherd's italics] to classical and academic psychology which, so far as I can detect, has little bearing on historical problems, but rather to psychoanalysis and its later developments and variations as included in the terms 'dynamic' or 'depth' psychology.[5]

Another advocate is Lloyd de Mause, writing in the *Journal of Psycho-History* (1975):

history cannot be science in any strict sense of the terms and history can never regard it as part of its task to establish laws. Written history may, in the course of its narrative, use some of the laws established by the various sciences, but its own task remains that of relating the essential sequence of historical action and, *qua* history, to tell what happened not why.

Psycho-history . . . is on the contrary specifically concerned with establishing laws and discovering causes. The relationship between history and psycho-history is parallel to the relationship between astrology and astronomy, or if that seems too pejorative between geology and physics. . . . Psycho-history [is] the science of historical motivation.[6]

Shepherd comments:

For de Mause 'psycho-history' is a wholly new discipline, 'less a division of history or psychology than a replacement for sociology, based on a set of problems, a conscious methodology and criteria of excellence all its own.' The contents of a typical issue of the *Journal of Psycho-History* illustrate the products of this ambitious claim . . . The self-explanatory titles of the articles are: (1) Kissinger: a Psycho-history; (2) Pedagogy as Intrusion: Teaching Values in Popular Primary Schools in Nineteenth Century America; (3) Infanticide in the Province of Canterbury during the Fifteenth Century: (4) Autobiography as a Key to Identity in the Progressive Era; (5) Psycho-history and Psycho-therapy; (6) Psychological Analysis and Presidential Personality: the case of Richard Nixon.[7]

Now, just as crude astrology has adherents who make claims that more serious astrologers would not support, so we have some non-professionals who advocate a crude psycho-history. But they are not necessarily journalists writing to cheer up the readers of a popular paper or magazine. Here we have Leo Abse MP writing, it seems seriously, about the working of Parliament as interpreted from a psycho-historical position:

our rules are drafted to seek to contain our perversions. The drive of our lust for dominance and power which has brought most of us to our places must be checked if the institution is to survive. The impress of the anal phase, the stage of infantile libidinal development when, accompanied by aggressive fantasies, the child is fascinated by the mastery of his body through sphincter control, is too deeply printed upon our characters. Although most members would fiercely resist any interpretations which hint at the origins of the nomenclature of our institutions and proceedings, the blunt fact is that we take our seats every day in a chamber where we are continuously passing motions.[8]

Professor Shepherd does not find these words as significant as does Mr Abse. There are many words in English with more than one meaning, and these meanings are neither etymologically connected nor conceptually connected except in so far as one word may be linked to another for bizarre effect as in punning. Examples are: bore, set, organ, train. Returning to the key words in the Abse quote: 'seat', 'chamber', 'motion'. It is true that they can all be connected so as to relate to the lust for power – but almost any word can be connected to another via intermediaries:

> 'Why is a penny a shilling?' 'A penny is a copper, a copper is a policeman, a policeman is a bob and a bob is a shilling.'

Analogously, MPs take their seats and a seat is connected with buttocks, the buttocks are near the anus, defecation is through the anus, this involves sphincter control and a use of power. Therefore parliamentary language reveals the innate lust for dominance and power. Reliance on tenuous connections, ambiguities, unjustified assumptions and so on, is a feature of pseudo-scientific argument and pseudo-explanations.

Let us look at another part of Shepherd's article, where he refers to Hugh Trevor-Roper's criticism of psycho-history. He quotes from Trevor-Roper's article 'Re-inventing Hitler' in the *Sunday Times* of 18 February 1973. This article was about the book *The Mind of Adolf Hitler*, written in 1973 by Dr Walter Langer, brother of William Langer:

> Professor Hugh Trevor-Roper, for example, was scarcely able to contain himself in his assessment of Dr Walter Langer's book on Hitler, a topic on which he is an acknowledged authority, and he took the opportunity to deliver a memorable broadside.

> . . . Psycho-history does not only rest on a defective philosophy, it is also vitiated by a defective method. Instead of proceeding from demonstrable steps, from fact to interpretation, from evidence to conclusion, psycho-historians move in the opposite direction. They deduce their facts from their theories; and this means, in effect, that facts are at the mercy of theory, selected and valued according to the consistence with theory, even invented for the sake of theory. The defect springs, I suggest,

from the very nature of psychoanalysis. For psychoanalysis, at best, is a means of therapy, not of investigation. Even if it should cure, it is not thereby shown to be true. It is not a means of establishing fact but a therapeutic myth.

By way of illustration Trevor-Roper singles out the following passage, in which Langer describes how, after completing his work, he visited a female colleague who asked him what he had been able to find out about Hitler's childhood:

'Without attempting to be orderly I related the material as it came to my mind but omitted any appraisal of its possible significance. She listened intently for a while and then interrupted me saying, "Now I know what his perversion is!" And', exclaims the doctor, 'to my utter amazement, she was right!' i.e. she agreed with him.

Trevor-Roper notes that, in this account,

Dr Hall had not heard a single fact about Hitler's sexual life. Her 'far-reaching insight' into his alleged perversion was not based on any empirical or verifiable evidence at all. It was intellectually deduced by the rules of psychoanalysis, from some casual statements about Hitler's childhood which Dr Langer had picked up from stale and random gossip repeated fifty years afterwards and thousands of miles away. And yet the agreement of two psycho-analysts interpreting the same rules in the same way made it 'right': they 'knew.'
    I treasure the record of that meeting: it so perfectly illustrates the parallel which I have suggested between psycho-analysts and witch-doctors. I can easily imagine two grave old demonologists chatting comfortably by a porcelain stove in 17th-century Württemberg or Bavaria and discussing, over a flagon of Bocksbeuttel, the case of a suspicious old lady in a remote village. On a few scraps of gossip, insignificant to the layman but full of meaning to the expert – the furtive comings and goings of a black cat, the wasting away of one of the parson's piglings, an alleged rustle in the chimney-stack – they both reach the same conclusion; and 'by God!' exclaims one to the other, slapping his thigh and pouring out another gurgling Pokal, 'you're right!'[9]

It is clear that Trevor-Roper is convinced that psycho-history is a pseudo-science offering spurious explanations, but, equally clearly, the Langer brothers do not have this opinion. Their differences are not a dispute about the facts, but a dispute as to the significance of the facts. Trevor-Roper considers that statements about Hitler's childhood, even if true, are far too casual to be of any value. On the basis of such vague facts it is, he thinks, impossible to explain the actions of historical figures, any more than facts casually picked up from village gossip serve to support a theory that an old woman is a witch. He has another objection to the psycho-historical method, namely that the facts are 'at the mercy of theory'. This is not quite the same as facts being theory-laden, although, as we saw in Chapters 3 and 4, a fact can be *modified* by an explanatory theory. However, to say that a fact is at the *mercy* of a theory is to say that the fact is *always* interpreted so as to support the theory. Hence the theory can never be shown to be wrong. As we have seen this is the characteristic of pseudo-scientific theories and spurious explanations. It is not that they are wrong, rather it is that they are so formulated and interpreted that they *cannot* be falsified.

We saw that there is disagreement as to the status of psycho-history and the explanations it offers. These differences are even more marked when we consider explanations of behaviour in terms of Freudian theory and explanations of the success of psychoanalysis. Unlike subjects such as astrology and psycho-history, no serious scientist can completely discount Freudian theory in helping us to explain our behaviour; it would also be hard to deny that psychoanalysis has some therapeutic value. What is in dispute is whether the explanations offered by psycho-analysts, that is the *causes* they suggest for neurotic and/or psychotic behaviour are genuine or spurious explanations. I take a quote from Popper's *Conjectures and Refutations*, where Popper argues that both Freud's and Adler's theories are so formulated as to be irrefutable and therefore they cannot be taken as genuine explanatory theories.

> The two psychoanalytic theories . . . were simply non-testable, irrefutable. There was no conceivable human behaviour which could contradict them. This does not mean that Freud and Adler were not seeing things correctly: I personally do not doubt that much of what they say is of considerable importance and may well play its part one day in a psychological science which is testable.

But it does not mean that those 'clinical observations' which analysts naively believe confirm their theory can do this any more than the daily confirmations which astrologers find in their practice.[10]

Popper makes it clear that he does not think the theories are useless, only that they are not as yet sufficiently well formulated to provide proper explanations. Since they can explain anything they are, at present, a source of merely spurious or pseudo-explanations. The psychologist H. J. Eysenck directs his criticisms primarily against the use made of Freudian theory to provide spurious explanations.

In the 1890s Freud and his colleague Breuer observed that when the sources of a neurotic patient's ideas and impulses were brought into consciousness by hypnosis, she or he showed improvement. Freud developed a technique of encouraging his patients to associate their ideas freely so that they said whatever occurred to them. He noted that there could be difficulties in making free associations and he developed the theory that painful experiences were repressed, that is held back from conscious awareness. He held that the psychoanalyst must encourage his patients to transfer emotional attachments to the doctor and to analyse the origin of their neurotic or psychotic feelings. This would relieve, if not cure, them.

Modern opinion is divided as to whether this is a genuine scientific theory. As quoted above Popper does not think that it is and neither does Eysenck; this is because they hold that the theory is so formulated that it can be compatible with any facts and indeed with complete failure to achieve a cure. Eysenck would contend that *even when* the patient does show some relief, there is no ground for supposing that this would not have been achieved by any sympathetic and reasonably detached attention. He points out that such attention gives as good results as those achieved by psychoanalysis.

It does not follow that we have to dismiss the theory as useless for it may be that a psychoanalyst can get better therapeutic results than an ordinary sympathetic friend (or doctor) and yet it could still be the case that it provided a spurious explanation of the success of the therapy. Eysenck's criticism is severe; he makes quite plain what he regards as the criteria whereby spurious theories are to be distinguished from genuine ones. In his review of Fisher and Greenberg's *The Scientific Credibility of Freud's Theories and Therapy*,[11] and in the subsequent debate,[12] Eysenck agrees that it is important to consider the value of Freud's theories since their influence on modern thought is considerable. He points out, however, that they have more effect on non-scientists than on those experimental psychologists working in the field. He also points out that most of the great philosophers of science (such as Popper) think that the theories are too equivocal to be tested, that is they are so formulated that they are 'barricaded against criticism'. For example, according to Freudian theory, dependent men should prefer women with large breasts but, should a dependent man prefer women with small breasts the theory is rescued by appeal to 'reaction formation'. Eysenck says that this is an example of 'one of those Freudian mechanisms which make it impossible to perform any proper test of his theory or to produce proof or disproof'. Another way in which the theory is protected is by ignoring alternative theories that suggest therapies, for example behaviour therapy, which give better clinical results and which can *also* explain what is observed.

In his review Eysenck says that there is no proper criticism of Freudian theories of human behaviour and personality; evidence quoted is not assessed so that the results of serious and absurd investigations are treated indiscriminately. Indeed he says that evidence which is in fact against Freudian theory is interpreted as though it were supporting it. He says that the reason why so many people believe that psychoanalysis works is that neurotics tend to get better anyway; this spontaneous remission has to be allowed for. Put succinctly, Eysenck is arguing that Freudian theories are so vaguely expressed that no firm consequences can be derived from them and they cannot therefore be regarded as giving explanations of behaviour.

Now, of course, these views are opposed by the authors of the book. Apart from appeals to Eysenck's known opposition to Freudian theories (these are irrelevant and tend to weaken their case), Eysenck's opponents say that the prediction that dependent men prefer large-breasted women cannot be derived from anything in the Freudian oral character theory. They say that they *did* consider alternative theories and that they do not think that behaviour therapy does give better clinical results. We may note that here, unlike the argument over psycho-history, we have an argument about certain facts. *This* makes the controversy less interesting philosophically since philosophers are primarily concerned with discussing the interpretation and the significance of facts and not with experimental investigation and clinical trials.

It may seem strange that Eysenck should begin his attack by asserting that Freudian theories are so vaguely expressed that they cannot be tested and then later state that they have been shown to be false. 'There is now I contend a very large body of work to indicate that the Freudian prediction is wrong.'[13] But his counter to this criticism would be that he objects not so much to the theory as formulated by Freud, as to the presentation and interpretation by its present-day protagonists. In his view it is they who have so 'developed' Freudian theory that it has become unfalsifiable and therefore a source of spurious explanations.

> an inventive and keen young psychologist performs an experiment testing a Freudian hypothesis; he disproves it; a Freudian spokesman then alleges (a) that the theory did not really make this particular prediction and (b) that the experiment was not really designed to test it.[14]

Here we are not concerned with who is right; that is a matter for psychologists and psychiatrists to decide. The point of studying this particular discussion is to show that the distinction between spurious and genuine explanation is not always clear-cut. We may say that astrology is almost certainly a source of spurious explanation, probably so is psycho-history, and that Freudian theories are suspect if loosely formulated. It may be that the human sciences tend to accommodate spurious or doubtful explanations. Perhaps there is something irreducibly 'fluid' about theories of human behaviour and therefore an irreducibly non-lawful element in any explanation of human behaviour. If this were so psychological theories would have at least an element of spuriousness because conclusive falsifiability would be difficult, perhaps impossible. It remains to be seen whether this particular feature really does distinguish the human from the physical sciences for falsification of any explanatory theory (even a physical theory) is not generally a simple matter.[15]

In this chapter we have been concerned with the concept of spurious explanations and the criteria which distinguish a genuine empirical explanation from a spurious one. These are: predictive power, falsifiability and acceptance by the scientific community especially as the best of one or more alternative explanatory theories. Some philosophers (e.g. Hempel), would take the first criterion as most important; others (e.g. Popper) would take the second; others would take the first two together – they go together in practice (e.g. Eysenck) and some would regard the third as most important (e.g. Thagard – see notes 2 and 3). In the next chapter we shall consider a rather different kind of explanation, one that appeals to goals or intentions rather than directly to laws.

10

Teleological Explanations

Teleological explanations are explanations in terms of actual or expected outcomes; the *explicans* states the actual or expected consequences of the *explicandum*. It is helpful to discuss teleological explanations in two categories: purposive explanations and functional explanations. The former attribute some *active* goal-directed behaviour to the *explicandum*-agent whereas the latter imply a more passive quality whereby certain consequences occur. The two formulae 'T' and 'F', given by Wright, illustrate this:

(T) S does B for the sake of G iff:*
  (i) B tends to bring about G.
  (ii) B occurs because (i.e. is brought about by the fact that) it tends to bring about G.[1]
(F) The function of X is Z iff:*
  (i) Z is a consequence (result) of X's being there,
  (ii) X is there because it does (results in) Z.[2]

* 'iff' signifies 'if and only if'.

As we shall see, the distinction between purposive and functional explanations is not clear-cut and, in addition, it may be masked by loose terminology. For we do sometimes allude to the function of a person: we may explain the presence of a nightwatchman by saying that his function is to protect property. In this type of explanation the individual is being treated as an inanimate object and, for this reason, an ascription of function can be, and can be intended to be, insulting: 'Your function is to guard this place, not to give opinions'. On the other hand we sometimes explain the behaviour of an inanimate entity by ascribing purpose: 'The thermostat aims to keep the room at 65°C'; 'The missile makes for the target'. The ascription is metaphorical (or there is implicit reference to a human designer), but it is made because we find a teleologically orientated explanation satisfying. The situation in regard to animate but non-human individuals is rather different since the ascription need not be metaphorical and an explanation may legitimately appeal to goal-directed behaviour even though there is no implication of a conscious intention or desire. Plants and animals are active as compared with inanimate objects but only some of the higher animals are likely to carry out any actions. Here the term 'action' will only be applied to some human behaviour, though this is not to deny that it may be applicable to certain kinds of animal behaviour – that is left open.[3] Here 'action' is used for human behaviour that is accompanied by conscious purpose so that actions are distinct from bodily happenings. The latter are internal bodily activities (heartbeat and respiratory movements, for example) that we do not normally consciously control. A considerable amount of overt behaviour, simple reflexes, conditioned reflexes and habits, may also be treated as happenings, for they are on the borderline between happenings and actions in that though they are customarily carried out without conscious thought we sometimes plan to act in accord with habit and also we can consciously control such actions if we attend to them.

Charles Taylor makes this point when discussing actions as *directed behaviour*:

> the borderline between the behaviour we call action and that to which we refuse the name is very ill-defined. And this in turn arises from the fact that there is no sharp demarcation between directed behaviour and that other range of the organism's movements which cannot be described in this way. . . . The scale runs from blinking, shivering and sneezing, through yawning and laughing, to fidgeting and doodling, then to walking, writing, speaking, where we come to behaviour which is virtually always directed.[4]

All teleological explanations, whether purposive or functional, appear to differ from explanations conforming to the D-N model in that they appeal to an event that is a goal or outcome and is therefore, necessarily, later than (after the occurrence of) the *explicandum*. For this reason many philosophers have been chary of accepting them; they distrust an explanation that appears to explain what has happened, or what is happening, by what will happen. They point out that such an explanation is not compatible with our view of causality which requires that the cause of an event shall precede or, at the latest accompany, that event. They contend that teleological explanations are fundamentally misconceived and are logically incoherent for it is not logically possible for an event subsequent to another to be the cause of that other. Hence they object to teleological explanations of evolved adaptations: for example the explanation that a moth has black wings so that it will be less easily detected on sooty trees and therefore more likely to survive. They object to teleological explanations of behaviour: for example the explanation that a rat runs through a maze in order to find food. They also object to teleological explanations of human actions: for example the explanation that X went to London in order to attend a meeting. These kinds of explanation are held to be misleading since consequences cannot cause events and therefore cannot explain events.

However, it is the criticism that is misleading for teleological explanations are not intended to suggest that the goals described in the *explicans* bring about or directly produce the event or phenomenon featured in the *explicandum*. The connection is via general quasi-laws that are not usually made explicit. Wright's formulae: 'B tends to bring about G' and 'Z is a consequence of X's being there' can, therefore, guide us to a cause-like *explicans*. Thus the first explanation can be set out as:

Evolution tends to favour the survival and reproduction of well-adapted individuals;
Butterflies with black wings are less likely to be detected by predators on sooty trees and are therefore better-adapted;
Therefore they will be most likely to survive and reproduce;
Therefore we are likely to see butterflies with black wings where there are sooty trees.

The second explanation can be set out in a similar way. The third, an explanation of a human action, is of a different type but it too is not open to causal criticism for here the implied cause is an *intention* (to attend a meeting) and this does precede the journey to London.[5]

In addition it must be stressed that a teleological explanation need not be offered as a substitute for a physical causal explanation; a teleological explanation does not exclude a physical explanation and is not incompatible with a physical explanation. The latter suggest efficient causes, the former show final causes; in non-Aristotelian terms we may say that the physical explanation suggests the means, the teleological explanation suggests ends.

However, some critics are not happy with explanations in terms of final causes; their objection to a teleological explanation is based not on the fact that these explanations appeal to consequences that happen *after* the event explained but rather on the fact that teleological explanations, especially of human actions, need not invoke a physical event as cause. In their view teleological explanations are at best makeshift, and *faute de mieux*, because they do not suggest a *real* cause which *must* be a physical event. These critics may be prepared to accept that appeals to consequences are not illogical but they object to there being no reference to physical laws; for since a genuine cause must, in their view, be a physical event a teleological explanation is inadequate and likely to be misleading. Moreover, in so far as they give (spurious) satisfaction they may be dangerous for they can discourage further inquiry.

These critics are metaphysical materialists for the view that the only genuine causes of events are physical changes entails the view that events occur only as a result of interaction between material entities. Consciousness has to be admitted but as a mere epiphenomenon (see below); matter is the sole ultimate reality. It is impossible to demonstrate that this, or any other metaphysical position, is right or wrong for metaphysical beliefs provide us with a framework whereby we order our experiences and therefore they cannot be used to test our experiences. All we can do is to compare the frameworks provided by different metaphysical beliefs and so indirectly assess those beliefs.

As stated above, physical explanations are not incompatible with teleological explanations and we do not have to commit ourselves to metaphysical materialism to agree that teleological explanations may be complemented by physical causal explanations. Indeed we can go further than this (without being committed to materialism) and acknowledge that at least some teleological explanations are underpinned by physical explanations and that, in some circumstances, the latter are genuinely more fundamental. It is possible that all functional explanations, in addition to some explanations of purposive behaviour, will be shown to be underpinned by more basic physical causal explanations, though even if this is the case, it would not follow that the less basic teleological explanations were redundant. I suggest that though some purposive behaviour may be thus underpinned there is also some which is unlikely to be able to be satisfactorily described or explained in terms of physical causes (preceding physical events) alone. It is, for instance, unlikely that we will find physical explanations of certain kinds of animal behaviour, particularly that of mammals similar to ourselves, which will be regarded as not only more fundamental but also as fully replacing a teleological explanation. In the seventeenth century many highly intelligent people, Descartes and Wren, for example, thought that animals were nothing more than automata with no inner consciousness: without desires and without the capacity to feel pain or to have any inner experiences. To hold such a view, even at that time, would seem to indicate that intellectual ratiocination should be restrained by common sense and that theory unrelated to observation can produce nonsense. For then, as now, the main evidence that animals are conscious (sentient) and do feel pain, hunger, fear and so on, is precisely the same as the evidence that other human beings (each one of us has *direct* experience of her own sensations) are conscious. To be fair to Descartes and Wren we have to concede that today we do have more reason to believe that animals are capable of feeling for, since the publication of Darwin's *Origin of Species*, we have come to appreciate that there is no clear distinction between human beings and other living creatures.

Hence it does not seem unreasonable to suppose that teleological explanations of at least some aspects of animal behaviour will be as valuable and as necessary as teleological explanations of some aspects of human behaviour. The merits of the belief that at least some animals are not only conscious (sentient) but also are aware that they are conscious and have *conscious intentions* and *consciously appreciated goals* is not under debate here. At present teleological explanation of much animal behaviour is undoubtedly helpful and it is certainly compatible with the concept of a family of living creatures with common ancestors. This can be acknowledged and accepted without begging the question against those who would deny that animals have conscious thought and are self-aware.

The status of purposive explanations of human actions could also be left undecided but we are on different ground here because although there remains room for doubt that animals are *self*-conscious we cannot, unless we regard our own self as unique, regard human beings as anything but self-conscious. That is something most of us would claim to *know* and we frequently make use of appeal to conscious intentions and to conscious purposes to explain human actions. In no way are these appeals to be taken metaphorically; indeed the significance of the term 'conscious awareness' depends on the fact that our own direct experience of self-awareness is the paradigm of consciousness. Materialist philosophers do not, of course, deny this; what they deny is that conscious thoughts *per se* have any affect on human actions. They are convinced that the 'real' or 'ultimate' explanation must be a physical causal explanation in terms of the physical events in each person's body, especially in the brain. They appeal to what is already known of neurological activity: the micro-physical events and chemical changes in the brain cells and throughout the nervous system. They point out that clinical experiments show that mental activity including conscious experiences is invariably correlated with such micro-events, and they claim that all human behaviour will eventually be shown to depend solely on physical events – external and internal. For them physical (efficient) causes are fundamental. Their case is put by Ryan, he calls them 'theoretical realists':

> The impact of theoretical realism . . . is that it makes the physiological processes seem more unequivocally the *basic* processes, since they fit, in a way that psychological processes do not, the picture of that well-ordered scientific corpus in which the processes analysed by physiology can be shown to depend on those analysed by chemistry, and these in turn on various physical processes. And for this reason, a good many psychologists would want to say at the very least that psychological explanations were not scientifically satisfactory until they could be backed by an account of the physiological mechanisms which showed us how the psychological processes could take place.[6]

Materialists claim that conscious thoughts are irrelevant; they are epiphenomena because the actions would have happened anyway, that is without there being any such conscious thoughts. Wright points out that to say 'It would have happened anyway' is too facile for what we (correctly) pick out as the cause of an action or of an event remains a cause even if the event or action was over-determined and would have occurred without the factor we had picked out as cause. For example it may be said that a man's death was caused by a road accident and later it may be found that at the moment of impact he had a heart attack. Either event would have killed him; it is not incorrect to say that he died as a result of the accident just because he also had a heart attack, or *vice versa*. Likewise a person may intend to take a walk and it may be possible to show that certain chemical and physical events occurred in her brain. Either event, the intention or the brain events may be cited as cause. However, it seems to me that Wright does not allow for the materialist who insists that physical events are the *only* causes; for her the existence of an intention does not *over*-determine the action because it could not cause the action. His other objection is more telling for he reminds us that the subjunctive '*would have happened*' is suspect since we have absolutely no evidence to substantiate it. We do not know if an action would be the same or even if something similar *would have happened* if there has been no conscious intention. The materialist contention that thoughts *per se* are not causally effective is based solely on the metaphysical assumption that physical causes alone can produce physical changes and cause physical events; this begs the question.

Metaphysical materialists must insist on physical causation and, if they are to be consistent, they must also support the thesis of physical determinism, the thesis that all human actions are the same as other physical events (happenings) so that human behaviour is determined by physical events operating in accordance with causal laws. Elsewhere[7] I have argued that if determinism and the view that physical causal explanations are basic is adopted there is need to reassess our concept of a person and, as Ryan says,[8] we would then have to see each other and ourselves as detached entities without moral responsibility. Nevertheless we can analyse purposive explanations of human behaviour without begging the question against the materialist position and without denying the relevance of physical causal explanations. Here purposive explanations will be treated as goal-directed explanations and it will be assumed that they offer genuine explanations of human actions, based on appeal to conscious intentions, desires and beliefs. Materialists need not take exception to this for the behaviour of animate beings can be regarded as being explained *as if* it were goal-directed, and human actions can be regarded *as if* they were guided by conscious intentions, desires and beliefs. the materialists' right to hold that such explanations are purely interim explanations is not thereby withdrawn.

As explained above, if we ascribe a goal or purpose to behaviour we do not necessarily imply consciousness, or conscious awareness of the goal or conscious intention to achieve it. But the goal must be something that we, as the observers, acknowledge as a goal. Hence it might be said that any explanation of human behaviour must be subjective and any teleological explanation of animal behaviour will be illegitimately anthropomorphic. Wright argues that this does not matter as long as the goal itself can be objectively established as something at which the behaviour does in fact aim. It does not matter, then, if that behaviour is described in anthropomorphic terms for:

> Teleological behaviour is not *simply* appropriate behaviour: it is appropriate *within a certain etiology*. Establishing the etiology is what is central to the teleological characterization. For this, all the etiologically operative property must be is, (a) objectively determinable, and (b) dependent upon the goal; it does not matter at all that the descriptions under which humans find it easiest to make these assessments happen to be anthropomorphic. All that matters is that they be intersubjectively testable; and this they assuredly are. There is nothing essentially subjective or mysterious about anthropomorphism of this sort.[9]

and

> The anthropomorphic analogy may *explain our insight* into a complex phenomenon. The *justification of our teleological characterization*, however, must be that directive organization is the best etiological analysis or account of the movements; what is evoking them must be their tendency to produce a certain result.[10]

Wright points out that the assumption of a goal is the assumption of an explanatory theory since it is not possible to 'observe' an intention directly. The theoretical nature of the assumption is well marked in the attribution of goals to lower animals and to plants:

> Theoretical terms, in this context, are taken to refer to things that are in some strong sense unobservable. Hence the *justification* for postulating them has to be that their existence *best explains* some

phenomenon or other. Similarly, the justification for saying a certain bit of behaviour is goal-directed has to be that the best explanation of the behaviour is that it is being brought about by its tendency to produce a certain result. Furthermore, in some cases the goal-directedness of behaviour is very difficult to detect (e.g. lower animals and plants). In these cases 'goal directedness' would simply *be* a theory term, being applied only after checking the consequences of its postulation.[11]

Purposive explanations of behaviour *depend* on showing that the behaviour is goal-directed, and if the goal *is* shown then *ipso facto* we have an explanation. But frequently the mere ascription of a goal would be regarded as an inadequate explanation of a human action for very often we seek to know the individual's reasons and beliefs and we seek to discover whether the goal was intended.

Wright has pointed out[12] that 'intention' and 'intentionally' are not always simply related, for 'to have an intention' implies prior thought whereas 'to act intentionally', though it can be equivalent 'to act in accordance with intention' may also be used to indicate that the behaviour was not accidental or not by mistake rather than that it was in accord with a prior intention. Thus we may brake our car intentionally (and *with* prior intention) as we approach a well-known dangerous corner whereas we may brake intentionally (but *without* prior intention) if a child darts into the road ahead. However, the fact that there are examples of behaviour which cannot be straightforwardly classified as actions (they have something of the reflex about them, see above, p. 86) need not affect an analysis of the teleological explanation of actions. If we think that the act was done with prior intention it will be explained by appeal to intention; if we think there was no prior intention it will be explained as a reflex or happening. As we shall see, on the assumption that intentions (to achieve a goal) may cause actions it is possible to set out a teleological explanation in something like a D-N form and to show that the explanation appeals to (quasi-) laws.

To say that an action, in order to *be* an action, must be carried out with conscious intention is to say that the concept of action is logically connected to the concept of intention. It is irrelevant that the intention be realised and it is even irrelevant that the action be unlikely to realise the intention, perhaps highly unlikely to do so. The essential feature is that the agent acts in the belief that her intention will be realised and her goal attained, or that it is likely to be attained or might be attained. There must be an intention for the agent to perform an action; this is a logical/conceptual requirement. Now since the time of Hume it has been acknowledged that a cause must be logically *independent of* its effect; therefore since an intention is logically *connected to* an action it would seem that it could not be the cause of an action. However, like the objection to teleological explanation based on the necessity for a cause to precede its effect, this objection to intentions as causes is based on a misunderstanding. It is based on a misunderstanding of the nature of intentional explanation and a failure to distinguish general concepts from particular instances. The concept of action *is* logically connected to the concept of intention – this is not to be denied – but the concept of effect is also logically related to the concept of cause. There cannot be an action without prior intention and there cannot be an effect without prior cause – Hume would have agreed – it is a matter of what the terms mean. But just as any particular physical event, say a stone thrown at a window, is not *logically* connected to a subsequent event, the breaking of the window, so any particular intention, say to post a letter, is not *logically* connected to a subsequent action, walking to the post box. It is only after the physical events are related through an explanation based on cause and effect that they become conceptually connected; likewise it is only after the intention is related to the action through a teleological explanation that *they* become conceptually connected. As Hume pointed out, causal relations are not, even then, like ordinary logical relations (and neither are teleological relations); in either case it is not *logically necessary* that the related events will both occur: the cause may not be followed by the effect, the intention may not be followed by the action. Both kinds of explanation, the physical causal and the teleological, suggest a connection of fact, *factual necessity* (see below); experience leads us to make a conceptual connection.

There are occasions when explanations appealing to intention are held to be inadequate; as indicated above, we may wish to know *why* the intention to act in a certain way arose – if, for example the goal was not attained and the action seemed likely to be ineffective. In these circumstances we take the action to be the result of a belief that it would realise the intention and if, as we generally do, we assume that the agent is rational, we will seek the reasons for that belief. On this view we can regard the reasons for the belief in the efficacy of the action as the cause of the belief. It is not necessary that others should think the reasons justify the belief, others may think them insufficient, even spurious, only the opinion of the agent is relevant. (The agent may, of course, be unaware of the true reasons for her belief but in so far as analyses of teleological explanation goes we may still appeal to *reasons*, though perhaps unconscious reasons, as causing beliefs.) On this view reasons cause beliefs, and beliefs give rise to actions that may realise intentions and achieve goals; hence, via beliefs, reasons can be regarded as causes or contributory causes of actions.

This view of reasons as causes has been discussed by Derden. He stresses that the *reasons* for doing something must be distinguished from actually doing it and also to be distinguished from the goal that the action is intended to achieve:

> Assume that an individual does 0 and has reasons for doing 0. In particular A's reasons for doing 0 are that he wanted E and believed that doing 0 was the best means of obtaining E. In these circumstances someone might ask: 'Why did A do 0?' . . .
>
> . . . when one gives A's reasons in response to the question 'Why did A do 0?', one has *not* given the reasons for why A did 0. But one *has* given the reasons for something. The question is: For what does one give reasons, when, in appearance at least it seems that one gives reasons for actions? The *prima facie* answer to this question . . . is that when one gives A's reasons for doing 0, one is actually giving A's reasons for *believing* that he should do 0, or A's reasons for *believing* that he will do or shall do 0, or A's reasons for *believing* that he wants to do 0. . . .
>
> It is my contention that we give reasons for what people believe they *should* do, and not reasons for what they do. If we *can* give reasons for what people do, then we can do so if and only if we can give reasons for what people believe they should do.[13]

Derden distinguishes reasons for a *belief* that an action would be effective (or belief that an action should be done to achieve a desired end, E) both from the desired end, E, and from the action itself. He says that the question 'Why did A do 0?' is ambiguous for we may be asking 'Why did A want E?' (because we take it for granted that A believes that doing 0 will achieve or might achieve E) or we may be asking 'Why did A believe that doing 0 would achieve (or be likely to achieve) E?'. In the latter case we are asking for the reasons for A's belief. He argues that the disagreement as to whether reasons can be

regarded as causes is due to the intermediary role of beliefs being
overlooked:

> It is almost as if both sides have committed a form of *post hoc* fallacy
> in that both sides conclude that because an action occurs after the
> reasons, it occurs *because* of the reasons. Then both parties argue
> about the 'because' in the expression 'because of the reasons'.
> When this occurs, that which interposes itself in some way between
> the reasons and the action is left out. This 'something' is A's belief
> that he should do 0.[14]

We can take desires as fundamental causes of intentions (these may
be followed by actions) bearing in mind that rational beings will have
reasons for believing that certain actions will achieve or may help to
achieve their desires. We can then explain actions by appeal to a D-N
model using teleological premises. Compare:

Physical D-N: All metals expand when heated.
               This piece of iron has been heated.
               Therefore this piece of iron is now longer.
Teleological D-N:

(a) Simple intentional: All letters posted in the box are
collected.

A wishes this letter to reach B.

Therefore A walks to the box and
posts the letter.

(b) Developed intentional: A wishes this letter to reach its
destination.

A remembers the times of
collection and believes it will be
collected today.

Therefore A intends to post the
letter in the box.

Therefore A walks to the box and
puts in the letter.

A's belief that he knows the times of collection is justified by his faith
in his memory, this is the reason for his belief. The question 'Why did
A walk to the post box?' may be best answered by 'A wanted to get a
letter off to B', or by 'A wanted to catch the post', or by 'A

remembered that the post would be collected from there today'. This last answer explains A's belief that he should go to *that* post box.

However, though the form of these teleological explanations is that of the D-N model, it is plain, whether or not 'beliefs' are introduced into the argument, that there is no firm deductive link from the premisses to the conclusion for intention does not invariably necessitate action. The road to hell is paved with good intentions whether based solely on desire or on desire supported by rational belief. A further pair of examples shows this more plainly:

A wishes to reduce his chance of getting lung cancer.
Therefore A intends to stop smoking.
Therefore A stops smoking.

and

A wishes to reduce his chance of getting lung cancer.
A knows the evidence linking smoking to lung cancer.
Therefore A intends to stop smoking.
Therefore A stops smoking.

Von Wright says that explanations of human actions, such as those given above, are in the form of *practical syllogisms* and he contrasts them with logical deductions, that is Aristotelian *proof syllogisms* (of the form 'All A is B, All B is C, Therefore All A is C'), because the practical syllogism is not a form of logical demonstration – the conclusion, as we have seen, does not follow from the premisses as a matter of necessity. However, he thinks that the practical syllogism is of great importance in explaining actions and that it provides an alternative to the D-N model used in the natural sciences:

Broadly speaking, what the subsumption-theoretic model is to causal explanation and explanation in the natural sciences, the practical syllogism is to teleological explanation and explanation in history and the social sciences.[15]

He implies that the subsumption-theoretic (covering law or D-N) model is based on the proof syllogism and carries a notion of necessity, the conclusion *must* follow from the premisses. But this is to go too far if it is to indicate that D-N explanations show what is logically inevitable. The distinction between physical causal and teleological explanations could rest only on the notion that physical laws carry some notion of *natural necessity* that established (quasi-) laws about beliefs, intentions and actions do not carry. As a matter of common sense this distinction is worth making but it is misleading to imply that inferences from premises that are physical laws are analogous to logical inferences. That is to down-grade Hume's cogent criticisms of causal inferences based on past experience, which was referred to in Chapter 6.[16]

Indeed Hume had an inverted form of von Wright's argument for he suggested that our awareness of the lack of necessary connection between intention and action (will and deed) should show us that there was an analogous 'gap' between physical cause and effect (operation of body):

> If we examine the operations of body, and the production of effects from their causes, we shall find that all our faculties can never carry us farther in our knowledge of this relation than barely to observe that particular objects are *constantly conjoined* together, and that the mind is carried, by a *customary transition*, from the appearance of one to the belief of the other. But though this conclusion concerning human ignorance be the result of the strictest scrutiny of this subject, men still entertain a strong propensity to believe that they penetrate farther into the powers of nature, and perceive something like a necessary connexion between the cause and the effect. When again they turn their reflections towards the operations of their own minds, and *feel* no such connexion of the motive and the action; they are thence apt to suppose, that there is a difference between the effects which result from material force, and those which arise from thought and intelligence. But being convinced that we know nothing farther of causation of any kind than merely the *constant conjunction* of objects, and the consequent *inference* of the mind from one to another, and finding that these two circumstances are universally allowed to have place in voluntary actions; we may be more easily led to own the same necessity common to all causes.[17]

and

> It would seem, indeed, that men begin at the wrong end of this question concerning liberty and necessity, when they enter upon

it by examining the faculties of the soul, the influence of the understanding, and the operations of the will. Let them first discuss a more simple question, namely the operations of body and of brute unintelligent matter; and try whether they can there form any ideas of causation and necessity, except that of a constant conjunction of objects, and subsequent inference of the mind from one to another. If these circumstances form, in reality, the whole of that necessity, which we conceive in matter, and if these circumstances be also universally acknowledged to take place in the operations of the mind, the dispute is at an end; at least, must be owned to be thenceforth merely verbal. But as long as we will rashly suppose, that we have some farther idea of necessity and causation in the operations of external objects; at the same time, that we can find nothing farther in the voluntary actions of the mind; there is no possibility of bringing the question to any determinate issue, while we proceed upon so erroneous a supposition.[18]

At the end of the day all our laws of nature and all our explanatory theories must rest on observation of what is, effectively, constant conjunction and, as was pointed out in Chapter 6, there is no logical necessity embodied in our empirical laws and no logical necessity in empirical explanations. We are more aware of this lack of necessity in the connection between volition (intention) and action because we do not observe constant conjunction; we know that we all have conflicting desires and that our will and our intention can vacillate. Nevertheless, as Hume pointed out, there is enough regularity in human behaviour to enable us to suggest reasonably reliable laws (and quasi-laws) that can be the basis of causal explanations and predictions.

Just as to ascribe purpose or goal is, in itself, to offer a purposeful explanation, so to ascribe function is to offer a functional explanation and though we may decide that functional explanations are underpinned by physical causal explanation, it does not follow that the functional explanations are or will be redundant. They imply a context, they put the *explicandum* into a background, and they contribute to answering the 'Wherefore?' latent in the search for explanation. Wright[19] points out that functional explanations suggest *one* reason for the *explicandum* which is likely to have been selected from several possible reasons. A functional explanation is likely to answer the question 'Why X as opposed to something else?'. In context the number of alternatives is likely to be limited but, without reference to context, it would be virtually impossible to rule out *all* possible alternatives in a non-trivial way.

In certain contexts functional explanations are the *kind* of explanation that will give the understanding required. It may be helpful to have functional explanations of thermometers, thermostats and telephones for example; likewise a bodily sensation such as pain may, in certain contexts, be best understood as having a warning function: 'by its tendency to produce action designed to separate the damaged part of the body from what is thought to cause the damage'.[20] There are many organic processes: sweating, digestion, growth, reproduction, respiration and so on that may be better explained in terms of function even though physico-chemical explanations are available and are to be preferred in other contexts. Moreover there are still very many bodily activities that can *only* be explained in terms of function. For example though we can now give much more detailed physical descriptions and offer more explicit accounts of the chemical processes occurring in living cells than could be offered at the beginning of the twentieth century we have not been able to eliminate functional explanations of cell activity, we have just taken them one stage back: we can offer physical descriptions of the changes occurring in cell division (mitosis and meiosis) and we can use these to explain many features of growth and inheritance, but we still cannot explain why the zygote divides and then differentiates as it is seen to do in physical terms. We can refer to genes and genetic coding by DNA and RNA yet, at the last, the explanation appeals to the function of these complex molecules. In any case, however advanced our physical science, functional explanations will not be redundant for, like the need for teleological explanation, they give understanding in terms of final cause.

In addition functional explanations can help us to understand the *explicandum* by showing it to be part of a coordinated system, the function (or functions) of each part being an activity that helps to sustain the system:

> for something to perform its function is for it to have certain effects on a containing system, which effects help sustain some activity or condition of the containing system.[21]

The study of system control is called *cybernetics*, and this includes the study of self-regulating mechanisms in systems – for example a mechanism, within the system, for maintaining that system at constant temperature. Such mechanisms act by responding to the state of the system by initiating events in the system; this is *feedback*. Thus if the human body temperature rises above 37°C (and the person is in good health) sweat is produced to help bring the temperature down; on the other hand if body temperature falls below the optimum sweat production is decreased or ceases and there may also be vaso-constriction and goose-flesh – reactions which reduce heat loss (though the goose flesh reflex is vestigial for us humans and has little effect). Sometimes control is by *negative feedback*, that is one part of the system reduces activity if the system is over-active, or increases activity if the system is under-active. Thus a thermostat will act to reduce the production of heat by the system if the temperature rises above a 'set' temperature and will act to increase heat production if the temperature falls. The result of cybernetic studies is to replace quasi-teleological explanations with physical explanations that do not appeal to purpose even metaphorically.

Cybernetics has also been applied to brain function (as well as the functions of other parts of the body) and materialists would of course argue that through cybernetics it will be possible to eliminate teleological explanations of actions – the brain will be revealed as the physical co-ordinator of a complex physical system. However, as we have seen, it is debatable if this could ever be the case. Nevertheless it is likely that cybernetics will make further major contributions to neurology; in addition it may help us in explaining some aspects of social behaviour. Von Wright says:

> In feedback there is a concatenation of two systems. . . . the primary and secondary systems. A certain amount of the primary system's causal efficiency is deflected into the secondary system so as to keep it 'informed' of the operations of the first. This inflow of information makes the cause-factor of the secondary system operative. The effect is fed back into the primary system and 'orders' a modification to take place in the operation of its cause-factor. This closes the chain of concatenated operations.[22]

Thus some of the heat in a heating system is deflected to the thermostat, which responds by changing the supply and/or the combustion of fuel and so modifies the heat (the cause-factor). Von

Wright would say that the input into the secondary system (the rise in temperature produced by the heat in our example) is 'information' which alters the thermostat (the cause-factor of the secondary system). That change is fed back into the primary system by making an 'order' on fuel supply and/or use and so modifies the cause-factor (the heat output) of the primary system. He continues:

> Now think of a case where the actions of agents can be said to steer a society in a certain direction through the decisions which they make effective, by building up 'normative pressure' and perhaps by the occasional use of such means as physical compulsion or violence. Assume that there is a part of the society which does not participate in the decision-making of the power group but which is informed about the results and enlightened enough to reflect about their consequences – . . . This insight . . . may then call forth in that other group a desire to influence the power group so as to give its efficacy a new direction or otherwise temper it. In the absence of institutionalized channels for communicating new directives to the power group, the 'feedbackers' will have to resort to forms of communicative action such as demonstrations, protests, strikes, sabotage, *etc.*, which are not regulated by the existing rules of the social game and are sometimes even contrary to these rules.[23]

Von Wright says that striking is an example of this: the employers, say the management of a nationalised industry such as the Coal Board, are the steering agents who establish normative pressures (rules) so as to reduce the labour force through early retirement, which may be made compulsory and perhaps be accompanied by redundancies. The employees foresee job loss and trade unionists (including those not directly concerned) foresee changes in labour relations. If there are no institutionalised channels, or if these are not effective, demonstrations and strikes may follow. For von Wright this is an example of a feedback mechanism: the secondary system, the union, being informed by, and acting on, the primary system, the coal industry and its management. But, as he points out, though the parallel can be helpful in explaining what happens, there is no question of replacing an account in terms of intention by a physical causal account. Unlike a feedback account of a thermostat or a sweat gland we do not substitute a physical account for a quasi-teleological one; rather we develop an account in terms of individual intentions to show how these can lead to group action.

Functional explanations of social activities may not always be reducible to the intentions of individuals. Von Wright claims that the activity may transcend individual purposes:

> A demonstration has a purpose which can somehow be 'extracted' from the purposes of individual men. But in what way it can be extracted is not easy to say. A folk festival or religious procession is only remotely, if at all, connected with purposes. Perhaps some people took part in the festival in order to amuse themselves. This would explain their presence on the occasion. But knowledge of their, and other participants', purpose in joining the crowd would not tell us that what is going on is a folk festival. . . .
>
> The answer to the question what is going on here is not a teleological explanation of the actions of individual men. It is a new second-order act of understanding.[24]

He wants to distinguish the 'How' and more importantly the 'Wherefore' questions '*What* is the event?' and 'What is the significance of the event?' (e.g. a folk festival, a religious procession or a Royal wedding) from the 'Why' question, 'Why does it happen?'. He points out that any given account will prompt a 'why' question, and the answer will demand a new interpretation and then a further 'why?'. There is a hierarchy of questions and answers:

> Something which used to be thought of as a reformatory movement in religion may with a deepened insight into its causes come to appear as 'essentially' a class struggle for land reform. With this reinterpretation of the facts a new impetus is given to explanation. From the study of the causes of religious dissent we may be led to an inquiry into the origin of social inequalities as a result, say, of changes in the methods of production in a society.
>
> With every new act of interpretation the facts at hand are colligated under a new concept. The facts, as it were, take on a 'quality' which they did not possess before.[25]

This is closely related to our original analysis of facts and theories in Chapter 3 (see p. 31); functional explanations of social behaviour exhibit the same inter-relationship of fact and theory as do physical causal explanations.

Perhaps the purpose and significance of social activities, rites and customs may be more readily interpreted and understood by studying the society as a functioning whole, as a system, rather than by trying to 'extract' social meanings from the personal purposes of the individuals who make up the social group in question (see quote[24] above). Harvey stresses the way functional explanations show the *explicandum* as part of a whole.

> Functionalism is . . . sometimes associated with a doctrine of 'wholeness'. . . . More recently the concepts of organic or functional wholes has been raised in a new and rather interesting form through the application of systems of thinking to geography.
> . . . there are numerous situations in which it is common to talk of . . . functional wholes. The distinctive feature of such situations is that the behaviour of some system is not determined by the individual elements within it, but rather that the behaviour of the individual elements is determined by the intrinsic nature of the system itself (i.e. the whole). The individual elements may thus show a high degree of mutual independence but to demonstrate this empirically is not to demonstrate that the whole determines the part. Some writers have suggested however, that any system that exhibits a high degree of internal organisation should be deemed as a functional whole.[26]

He points out that a system is an abstraction, it assumes a self-contained closed unit. All the same, in geography, it is a valuable guide to explanation. He also points out that one system may be embedded in another system:

> A car may be an element in the traffic system, but it may also be regarded as constituting a system. . . . there are two ways in which we can conceive of an element in some level in the hierarchy of systems. It can be regarded as an indivisible unit that acts as a unit (e.g. we may think of a firm as deciding or responding) or it may be regarded as some loose configuration of lower order elements (e.g. individuals within an organisation interact with individuals in other organisations).[27]

The system as a system is regarded as closed (see above) but as a subsystem it is acted on from outside:

> In general terms the *environment* of some system may be thought of as everything there is. But it is useful to develop a much more restricted definition of an environment as a higher-order system of which the system being examined is a part and changes in whose elements will bring about direct changes in the values of the elements contained in the system under examination.[28]

Thus the values implied by the high-order system can introduce or modify values of a lower-order system rather as the language of a higher-order physical theory can modify the language of a description which is supplied in the language of a lower-order theory (see also Chapters 3 and 4). To go from a lower-order system of a car to a higher-order system of traffic implies different interests and is likely to carry new value considerations, for example a dislike of traffic congestion, of pollution by exhaust or lead poisoning. This can modify the original description of the lower-order system, that is of the car. Similarly to go from a lower-order system of a farm to a higher-order system of the economy and perhaps from there to an even higher-order system of the biosphere entails changes of interest and introduces values that can affect the description of the farm.

Harvey considers that:

> The methodological strength of functionalism really lies in its emphasis upon inter-relatedness, interaction, feedback, and so on, in complex organisational structures and systems. This kind of approach to problems has been extremely rewarding where it has been freed from metaphysical connotations. It thus seems very valuable to enquire how central places function in an economy, simply because it opens up a whole range of questions which, if we can provide reasonable answers, will yield us deeper understanding and greater control over the phenomena we are examining. The world is manifestly a complicated place, and the heuristic strength of functionalism is that it directs our attention to this complexity. . . . As a methodology it provides us with a series of excellent working assumptions about interactions within complex systems.[29]

The metaphysical connotations to which Harvey refers are the dangers of an approach based on the assumption that a system cannot be explained other than by taking the interaction of its parts as serving some function and that 'the whole' as a unique entity has a purpose. Neither do we have to take the *apparent* function as the best methodological tool; it is possible, and generally helpful, to distinguish the apparent or manifest function from a more profound latent function. Thus the apparent function of a central place, such as a market, is to distribute goods and services in an efficient manner. But more penetrating study may show that the more important, though latent, function is to satisfy the needs of people to meet and to talk to each other.

Harvey's view of functional explanation is that it is relatively weak because it does not provide conclusive causal (in the physical sense) explanation. In the quotation below he refers to Hempel, and this suggests that, like Hempel, Harvey takes the D-N model as ideal. But if we take the D-N model to represent only one aspect of explanation then functional explanations can be regarded as complementary; Harvey appears to allow this:

> Functional analysis can provide, therefore, only a weak kind of explanation by showing that at least one out of the range of traits must exist to fulfil a particular need. This is a rather unilluminating conclusion, for it amounts to stating that goods and services must be distributed in an economy *somehow*, for otherwise the economy will not function. Hempel suggests that the appeal of functional analysis is 'partly due to the benefit of hindsight', in that we already know that markets do exist. Functional analysis provides a convenient method of identifying some of the necessary conditions for explaining a given trait, such as the occurrence of central places.[30]

Ryan is more critical than Harvey; he accepts that there are valid and useful functional explanations of artifacts and of organs but he has doubts about functional explanations for the social sciences. He is not concerned with possible failure to conform to the D-N model; rather he thinks that functional explanations are misleading because they conceal the fact that it is the desires, beliefs and intentions of individual human beings that account for, and therefore should explain, social customs and events. In this view it is teleological explanations of the actions of individuals that are needed; he points out that in proposing an explanation of economic affairs:

> no one would dream of employing functional explanations – we do not, for instance, believe that a price drops in order to clear the market of a glut; we know that the mechanism at work is one which involves the wishes of sellers to get something for their wares, and it is on this that our explanations are founded.[31]

Trigg points out[32] that if social scientists can show uniquely social structures then sociology will be an independent basic discipline, whereas if social behaviour can be more adequately explained in terms of the various thoughts, beliefs and intentions of the individual members then psychology will be of more fundamental importance. Can the workings and the influences of such institutions as the Roman Catholic Church, the United Nations, trade unions and the politbureau be explained most satisfactorily in terms of the thoughts, intentions and beliefs of the people who produced them and who maintain them? Have they, as Trigg says, an ontological status apart from the activity of those producing them?[33] He thinks that it is plausible to deny (as Ryan would deny, see above) that these kinds of human creations can exist independently of individuals and their ideas though he realises that it is then necessary to explain why certain social customs and attitudes, the incest taboo, for example, appear to be universal. He suggests that this could be due to our common *human* nature, if we want to stress the importance of each individual's thoughts (consciousness) there is less place for such a general notion as human nature. Indeed human nature can be rated as a product of conscious awareness and human nature should not be regarded as an immutable genetic 'given'.[34]

Should we then dismiss functional explanations of human social behaviour and insist that teleological explanations must be in terms of the desires, beliefs and intentions of individuals? I suggest that despite the danger of being misled, there is a place for functional explanations. We have already seen that they 'place' the *explicandum* into a system, and this may be helpful in a social context. In addition a functional explanation will remind us that there is an instinctive and non-rational side to human nature and that is revealed in our behaviour. Though human actions are of logical necessity related to intentions and beliefs we do not have to assume that all intentions are what the agent thinks and *a fortiori* we do not have to assume that all human social behaviour consists of actions. A comparison with the social behaviour of animals is illuminating: the social behaviour of ants and bees in defending their nests and hives, in spreading information about sources of food, in migrating; the courtship ritual of many animals and the marking of territory. We take this to be instinctive and we describe it in such a way as to invite functional explanation. This kind of instinctive behaviour is held to have evolved through natural selection and we may surmise that at least some human social behaviour has similarly evolved. That behaviour can usually be explained by appeal to individual desires and overt (or covert) intentions *post hoc*, but this may be a spurious rationalisation. Functional explanations of, say, the existence of social classes (pecking order), the status of women (sex and courtship), the treatment of children (rearing of the young), treatment of the sick and the elderly (preservation of the herd) and the nature of religious ceremonies and leadership (ritual, magic and dominance) may be more illuminating and more accurate, than explanations based on individual intentions. If we wish to understand social customs and attitudes and if we wish to change them, we shall be in a better position if we acknowledge the instinctive and non-rational aspects of our behaviour.

11

The D-N Model and the Human Sciences

In the previous chapter it was shown that, at least for the present, teleological explanations are necessary for interpreting much of human behaviour. We need to know individuals' desires, beliefs, intentions and goals in order to explain their actions satisfactorily; in the fields of social sciences, studying the behaviour of communities, we look for purposive explanations in terms of individual aims but we also seek functional explanations of social activities.

It was shown that teleological explanations (both purposive and functional) might depend on an explicit or implicit appeal to laws (and quasi-laws) and that they could be set out as D-N arguments. Hence it would seem that the study of social behaviour, as carried on within the social sciences, could be analogous in the study of inanimate objects as carried on within the physical sciences. Investigators in both fields seek causal explanations and expect to provide grounds for reliable predictions. However, there are those who argue that it is not possible to provide causal explanations of human behaviour (individual and social) that are analogous to causal explanations of physical events. There are, they say, both factual and conceptual objections.

The factual objections are based on the observed differences between human actions and activities and the behaviour of inanimate objects; these are:[1]

(1) The human sciences deal with such complex data that it is

impossible to specify the initial conditions accurately enough to give worthwhile minor premisses for the D-N argument.

(2) The laws of human sciences (potential major premisses of D-N arguments) cannot be as firmly established as the laws of physical science. The latter are obtained largely from controlled experiments; they are conducted in isolated and somewhat artificial conditions but in physical fields this is relatively unimportant and hence experimental results have practical application; in addition the experiments are (and must be) repeatable. But it is, in general, impossible to set up experiments in the investigations in human sciences: for practical reasons (as in history) or for moral reasons (as in psychology). In the few cases where experiments can be carried out the isolation and the artificiality of the conditions makes the results of little practical significance. In addition it is rare that experiments can be repeated and so independently confirmed.

(3) Generalisations about people are about relatively small numbers of individuals whereas generalisations in the physical sciences almost always cover large, sometimes indefinitely large, numbers of entities. For example, one litre of gas at atmospheric pressure and room temperature contains millions of times more molecules than there are people in the world.

(4) Inanimate objects (and non-self-aware animate entities) have no conscious purpose and make no choices whereas people are able to exercise choice in many situations; they act as though they are not determined by physical laws and so differ from objects.[2]

(5) Predictions made about the behaviour of inanimate objects (and non-self-conscious entities) do not affect subsequent behaviour, whereas predictions of people's actions can and do affect their subsequent actions; there is therefore an inescapable reflexivity in explanations offered in the human sciences.

(6) Individuals are unique and certain individuals can and do have a disproportionate influence on events.

These objections have been answered as follows:

(1) Admittedly there are practical difficulties in knowing all the

relevant initial conditions, and it must be conceded that each initial situation is likely to be unique; but it is probably true to say that even in the fields of physical sciences we cannot be sure we know all the relevant initial conditions. There is a difference but it is only a matter of degree. It does not follow that in human-science fields we know so little that we have no minor premises for a reasonable D-N explanation.

(2) By no means all of the physical sciences use evidence from laboratory experiments to establish laws: geology, astronomy, metereology, for example, are sciences where there can be little controlled experimentation. Experimental sciences such as physics and chemistry use evidence to establish laws applying to ideal models; though, as we saw in Chapter 5, it happens that the ideal models of physical science, for example the ideal gas, behave more like real entities (actual gases) than do the models of the human sciences, for example, 'rational man' and 'working class'. Yet this does not reflect a fundamental difference, it is a matter of difference of degree. The same applies to confirmation obtained by repeating experiments; admittedly in those physical sciences where experiments are possible they (generally) are less difficult to repeat, but even purely physical conditions can never be precisely the same. To the extent that experiments are possible in the human sciences there can (generally) be repetition and where this is not possible, as in history, critical observation of analogous situations can give some degree of confirmation (or falsification) of a suggested law or quasi-law. Thus these regularities are subject to some test, not so rigorous as the tests of physical science but rigorous enough for them to be taken as empirical laws and as premises for a D-N argument.

(3) We do not necessarily need large numbers to make a valid generalisation:

> a law of some intellectual development might be concerned with not more than a hundred or even as few as ten philosophers or authors. Provided all authors concerned are taken into account, investigation of macro-laws is still justified, since the authors have been singled out from considerably larger groups. Several years ago the statistician Bortkiewicz obtained his 'law of small numbers' by studying soldiers of the German army who had died from being kicked

by a horse. We do not insist on the correctness of this special law of Bortkiewicz,* but the method in general is justified. E.g. a law of intellectual history would be fairly well founded, though it is based on the observation of only fifty French philosophers, those having been singled out from forty million Frenchmen.[3]

(4) We have already noted that causal explanation of the behaviour of inanimate objects depends on appeal to causal laws (Chapter 6) and that human actions can be explained by appeals to intentions and to beliefs based on reasons. We have seen that there *are* laws (or quasi-laws) of human behaviour. It does *not* follow that because people can choose human behaviour is completely lawless; we do *not* have to be physical determinists to rely on some regularities. The fact that we speak of people acting 'out of character' and not conforming to accepted codes of conduct indicates that we do have expectations, based on quasi-laws, that people generally do act in accordance with what past experience tells us is 'in character' and that they do tend to conform to accepted codes. Flew does not use the term 'quasi-law' but says:

> though I am saying that there are not and could not be true descriptive laws of nature explaining human actions, to say this is not to assert that actions cannot conform to regular – even predictable – patterns.[4]

(5) Predictions of future actions are not *necessarily* invalidated if the individual(s) concerned is (are) aware of the prediction. Where reflexivity does occur, for example the reactions of voters to the published results of opinion polls, the effect can be at least partly allowed for by treating reflexivity as one of the initial conditions. This does not necessarily lead to an infinite regress of premisses. Clearly, where people are unaware that they are being observed and, of course, unaware that their behaviour has been predicted, there is no question of reflexivity. Many medical tests designed to inquire into the efficacy and action of a new drug are carried out using two groups of people: one group receiving the drug and the other

* It asserts that dispersion is nearly 'normal' in such exceedingly small groups. Cf. L. v. Bortkiewicz, *Da Gesetz der kleinem Zahlen* (Leipzig, 1898).

group a placebo. In addition, a possibility of the doctor influencing the patients' reactions is eliminated by ensuring that those administering treatment do not know which group are getting the drug and which the placebo. However, it must be admitted that investigations and predictions cannot always be concealed and their effects cannot always be allowed for. We know that human beings may act differently if they are aware that they are being observed and that they will be affected by explanations of and predictions about their behaviour. As Ryan says:

> This, then, is one of the ways in which long-run predictions will run into trouble; and it must be noted that it is not analogous to anything at all in the case of the natural sciences. . . . the reason why the prediction itself alters the future circumstances depends on people coming to understand why that prediction would have been successful . . . the prediction alters their subsequent behaviour, not in virtue of an independent status as a causal antecedent, but because people understand the prediction, and it then affords them reasons for behaving differently.[5]

(6) It is true that each human being is unique, but we must bear in mind that *all* objects and events are unique; as Zilsel says:

> He who has ever worked in a laboratory knows that even every apparatus, if it is somewhat more complex, has its individual characteristics and has to be handled with its own special tricks. No two reflectors of the same brand are perfectly alike and even less two planets. In natural science the variety of objects is mastered by the method of gradual approximation; objects may be handled as analogous in the first approximation; their differences are taken into account in second and third approximations, when they are out together and compared in new groups. Variety of historico-sociological phenomena surpasses variety of objects in degree only.[6]

There was a brief reference to the objections to the D-N argument as a model for the human-science explanation in Chapter 6. This more extended consideration of objections and counters must lead us to conclude that though a D-N explanation of human behaviour cannot be as forceful as in the fields of physical science, it has a place in the human sciences; it will help in answering the question 'Why?', and it will be a basis for rational predictions.

But we now have to consider objections to the D-N model that are based on conceptual grounds. Winch, for example, says that it is logically impossible for there to be D-N causal explanations in the human sciences. In *The Idea of a Social Science and its Relation to Philosophy* he takes social sciences to be concerned with the study of human actions in various social contexts and he says that actions must be meaningful to the agents and to other members of their society. The meaningfulness or significance of an action can be related to the agent's intention and Winch thinks that both the intention and the significance depend on the agent being a member of a social group in which all the individuals acknowledge and accept certain rules, certain codes, of behaviour. Winch holds that there cannot be *private* rules of behaviour, that is rules made and conformed to by just one person, and so there cannot be actions that are meaningful just for the agent. His argument parallels Wittgenstein's argument against the possibility of there being a private language:[7] it must be possible to test both rules and language by reference to other people. Winch says that we need to look for the rules that govern behaviour if we are to find explanations of it and he believes that appeals to causal laws of behaviour are logically incoherent. For, he says, causal laws must be based on observed regularities, that is on a pattern of like actions in like situations, and actions and situations are only understood as being alike within a framework of acknowledged social rules. For example, a man raising his hat to a lady, and standing back so that she can precede him through a door, are seen as similar acts of gallantry within a certain set of rules, a certain code of social behaviour. The *cause* of the action is the desire to show special attention to a woman *in certain contexts* but the cause operates as a cause only within the framework provided by the rules of behaviour. Hence Winch contends that it is *rules* that underpin causal laws and those rules cannot then be used to explain the laws they support. In the example the actions are what they are because the rules are known by both the man and the woman; they govern the behaviour and cannot be explained by any causal laws of regular association that might be suggested by an observer. Winch says:

A regularity or uniformity is the constant recurrence of the same kind of event on the same kind of occasion; hence statements of uniformities presuppose judgements of identity. . . . criteria of identity are necessarily relative to some rule: with the corollary that two events which count as qualitatively similar from the point of view of one rule would count as different from the point of view of another. So to investigate the type of regularity studied in a given kind of enquiry is to examine the nature of the rule according to which judgements of identity are made in that enquiry. Such judgements are intelligible only relatively to a given mode of human behaviour, governed by its own rules.[8]

Winch's argument is opposed by Ryan, who points out that although an action may only be meaningful if it is understood as being in accordance with a rule or as part of a code of behaviour, this is but to say that an agent has *reasons* for acting in a certain way. As we saw in Chapter 10, reasons can count as causes of beliefs and therefore contribute to intentions and actions. In the example above, the man's actions are understood by him, the woman and other members of their society; he makes use of them to fulfil his intention to be gallant or polite. Because others may have similar intentions regularities in behaviour can be observed; agreed the regularities do not explain the rules, rather the rules produce the regularities, but it is not logically incoherent to offer a causal explanation of actions made meaningful through rules since these rules provide the reasons for the agents deciding to act as they do. Ryan says:

The importance of causal histories remains as great as ever, once it is recognised that in human action we usually employ the concept of cause as equivalent to that of having good reasons: we still want to follow events made intelligible by showing how under the circumstances one thing rather than another was the thing to do – and thus the thing to be expected. The importance of deductive relationships is as great as ever; only if we know that a rule covers all of a given class of cases, and that all of a given group of persons follow that rule can we move towards deductively certified prediction of their actions; equally, the importance of consistency and inconsistency are undiminished, when we consider that neither we nor the people we study can persist in following what are seen to be inconsistent rules. . . . The logical properties and the logical presuppositions made by explanations thus seem to be unchanged, even if the social sciences aim at the elucidation of rules rather than the establishment of regularities only, and at the unravelling of conceptual rather than contingent relationships. To this extent we can say that the form of the enquiry may remain unchanged even when its content is seen to be so radically different.[9]

But in any case does accepted social behaviour necessarily *presuppose* rules? There is a distinction here between social rules and rules of a game. The playing of a game does presuppose rules and we might come to learn the rules of a game, say cricket, by watching people play rather than being told the rules; indeed it might be easier to appreciate the significance of some of the rules by watching them applied in play. But it is also the case that people might *claim* to be playing cricket and yet not be following the rules, and in that case their claim would be false for they would not be playing cricket. The rules of cricket, or any game, are not dependent on what people actually do, the rules come first, they *determine* the game. For example people who say they are playing tennis are usually playing lawn tennis; 'real tennis' has very different rules and despite the fact that it is rarely played it remains *real* tennis. The schoolboy at Rugby who, during a game of football, ran holding the ball ('with noble disregard for the rules of the game') was no longer playing football, he had initiated another game. Likewise contract bridge is a different game from auction bridge because it has different rules. If you want to play a particular game it is *logically necessary* to obey the rules for that game; certain games (Canasta and Hop-scotch) may have fluid rules and hence will allow a range of behaviour but this must still be in accord with whatever rules there are.

Rules for social behaviour are not like this; the influence is more complex for there is an interaction between the rules (what members of the social group regard as acceptable or expected behaviour) and how people do in fact behave. Whether we observe an alien society or our own we will incline to the view that whatever the actors *say*, whatever the expressed code, the real rules are revealed in what they *do*. If we agree with Ryan that people follow rules for reasons they must understand the significance of those rules in order to *decide* that they will comply. They may so decide because they know that others expect them to comply and they wish to co-operate and be accepted in the group, and/or because they want to support the rule and thereby support the stability of the society, and/or because they think the behaviour demanded is commendable in itself. The man who takes off his hat to a woman may do so because he knows it is what she expects (and what others expect) and/or because he thinks that conventional politeness is socially desirable; he is unlikely to think that the convention is commendable (or disgraceful) in itself. That rule relates to highly conventional behaviour and we can readily suggest reasons why a man might decide *not* to conform. But we can see analogous factors inclining people to conform to a more important set of rules, for example those governing the institution of marriage. Until recently people might decide to get married (as opposed to living together unmarried) because they wanted to be accepted by other members of society and/or because they thought that marriage as a social institution should be supported and/or because they thought it commendable to enter into legal and formal moral commitments to another person. It may be that some people simply follow social rules 'because they are there', in such instances the behaviour does presuppose rules, but this does not have to be the case. Of course, as we see above, reasons *may* presuppose rules, as they do when we *decide* to play a game but they do not have to do so. As Trigg points out, this is particularly evident when the view of acceptable behaviour is changing so that the current rules lose their force. In effect individuals cease to conform as a matter of course and the rules change:

> If enough members of a society begin to treat marriage as a temporary legal arrangement rather than a solemn undertaking broken only by death, it may well be that the institution of marriage will change. Saying this, though, suggests that institutions do depend on the understanding of their participants to such an extent that, if the latter changes, then eventually so will the former. The rules will be changed to take account of the changed understanding and thus cannot be thought logically prior to that understanding.[10]

As was indicated in the last chapter, not all explanations of social (or of individual) behaviour rest on appeals to social rules and codes: they may appeal to observed regularities of human behaviour that appear in situations of stress, alarm, fear, anger, pity, frustration and so on and may best be accounted for by a functional explanation (that can also be put in a D-N form). Such behaviour is non-rational in that it does not involve deliberate choice or conscious (rational) decision but it is likely to exhibit enough regularity to be the basis of laws or quasi-laws. Lastly there may be interactions between individuals and inanimate objects (the swimmer in the sea for example) that show enough regularity to provide premises for physical causal explanations.

Causal explanations of behaviour can lead us to assess that behaviour from a particular viewpoint. As we saw in Chapter 10, p. 99, a functional explanation may answer the question 'Why X, as opposed to something else?'; a causal explanation may answer the question, 'What was the key cause in bringing about Z?'. Nagel points out that a causal explanation of an event may be a matter of 'picking out' some prior event or prior set of events that are held to be the most important factor in a given situation because of the particular interests of those seeking explanation.[11] He says:

> we often do make such claims as that broken homes constitute a more important cause of juvenile delinquency than does poverty, or that lack of a trained labor force is a more fundamental cause of the backward state of an economy than lack of natural resources. Many people might be willing to admit that the *truth* of such statements is debatable but few would be willing to grant that they are totally without *meaning* so that anyone who asserts them is invariably uttering nonsense.[12]

He cites various examples of situations that guide us in picking out a cause. For example there may be two factors: xenophobia and a need for markets, that cause a country to adopt an imperialist policy; but if the first factor remains constant whereas the need for markets increases, then that need may be cited as *the* cause of the imperialist policy. Again, a supply of coal and a trained labour force may be necessary for productivity, but if alteration in the labour force produces a greater effect on productivity than variation of the coal supply, then the former will be held to explain (i.e. be the cause of) any observed variation in productivity. Now these examples show that the values and interests of those seeking causal explanations can decide what explanation they offer and this shows just as clearly when we consider possible causal explanations of events involving single individuals. For example a car accident may be the result of a number of causes such as careless driving, tiredness, drink, mechanical defect, poor road, lack of road signs and bad weather but, in regard to the *same* accident, different observers may 'pick out' a single cause as the 'key cause' and what they 'pick out' will be determined by their values and interests.

Nagel treats such causal explanations as being based on quasi-laws that are the basis of expectations. But he thinks that they are better (and perhaps more often) taken as providing *post hoc* explanations rather than predictions. The status of quasi-laws has been discussed in Chapter 8 and it was shown that they cannot be regarded as having the same force as physical laws. They are best used by those who are familiar with the kinds of circumstances that can modify them; as Rescher says, the role of experts is crucial when it comes to the application of quasi-laws. The expert does not just know the significance of the concepts (experts in both physical and social sciences must have full grasp of the relevant concepts to *be experts*) but in the social sciences the expert must be able to assess the relevance of ill-defined circumstances and background and fringe factors that affect the application of any quasi-law. This sort of knowledge cannot be precisely specified; it is simply a matter of fact that those people who are familiar with the field are more likely to produce satisfactory explanations and are less likely to make false predictions, or rouse unjustified expectations. Rescher says:

> the expert has at his ready disposal a large store of (mostly inarticulated) background knowledge, a refined sensitivity to its relevance, through the intuitive application of which he is often able to produce trustworthy personal probabilities regarding hypotheses in his area of expertness.[13]

Rescher takes as examples medical diagnosis, investment forecasting and predictions as to the number of prints to be made for a movie, and he says that examples can be found in politics and in applied sciences such as engineering. The expert knows what laws (and quasi-laws) are relevant and also knows the accepted statistical probabilities but he or she is *also* able to judge the effect of a multitude of different factors that bear on the given situation. This is what Rescher calls 'intrinsic expertise' involved not only in applying a law or theory but also in denying the relevance of and therefore *not applying* a law or theory.

Now if we conclude that both conceptual and factual objections are not strong enough to make us exclude D-N causal explanations in the human (including social) sciences we must take especial care that any law (or quasi-law) serving as a premiss is not so vague or so general that it verges on the spurious or trivial. It must be tightened up by making it more specific and by defining the terms more clearly. Dray, in his *Laws and Explanation in History*,[14] takes as an example of an unsatisfactory law 'Rulers who pursue policies detrimental to their subjects' interests become unpopular'. Let us consider this as featuring as a major premiss in a D-N argument explaining the unpopularity of Louis XIV. It could be 'tightened up' by making 'policies detrimental to the subjects' interest' less vague (more specific): 'Rulers who involve their subjects in foreign wars and who maintain expensive parasitic courts become unpopular' for instance. But then it could be argued that such a premiss is misleading for the unpopularity was due to policies being unsuccessful (the wars not being won, the court not being splendid) rather than to the policies themselves. So the law should be further modified: 'Rulers who involve their subjects in unsuccessful foreign war and who maintain expensive but not splendid parasitic courts become unpopular'. But then it could be argued that what was amiss was that the unpopular rulers did not distract the subjects with 'appeasing policies' of 'bread and circuses' – as I write (January 1986) there are topical equivalents that come to mind. So another qualification can be introduced; indeed an indefinitely long series of qualifications and specifications may be introduced to make the law acceptable. What will happen, argues Dray, is that what started as a general law will eventually be reduced to a law that applies just to the situation, the *explicandum* (Louis XIV or a modern prime minister); we lose the major premiss. The 'law' becomes a specific assertion, nothing more.

We have to accept that it can be difficult, sometimes perhaps impossible, to establish worthwhile regularities that can provide a basis for a D-N causal explanation. We have to concede something to factual objections (1) and (2) after all; there is difficulty in establishing laws and quasi-laws relating to complex situations and of specifying initial conditions but, as we saw in answer (2) (p. 111), critical observation can give some basis for a genuine causal law or quasi-law. In addition, as Ryan[15] says, we can place some reliance on trends, especially in the short term, even though we are unable to find an explanation of those trends. The situation is analogous to the use of physical laws (or established universal generalisations and statistical generalisations) that are not yet supported by an explanatory, higher-level theory.

Von Wright thinks that causal explanations in history (and social science) have a place, but that it is subsidiary to wider explanations – explanations that we have called 'wherefore' explanations. For example, if a ruined city is excavated archaeologists may wish to know why it was destroyed, whether by earthquake, flood, conquest or some other catastrophe. But von Wright suggests that the *primary* interest is not in the bare fact that it was destroyed by this means or that but rather in the political, economic and cultural effects of the destruction and, if it was destroyed by men, in the reasons for this:

> causal explanations . . . are not *directly* relevant to historical and social research. . . . But they may be indirectly relevant in two typical ways. One is when their *explananda* have interesting 'effects' on subsequent human affairs. The other is when their *explanantia* have interesting 'causes' in antecedent human actions and conditions. . . . Thus, for example, if the destruction of the city was an act of envy or revenge on the part of a neighbouring city and if the destruction in turn became an economic disaster for the entire region, we have established a link between *the rivalry of the two cities* and subsequent *changes in the economic life* of the region. *This* is the kind of connection the historiographer is interested in.[16]

Von Wright says that historians will also be interested in explanations showing how something was possible: *how* Stonehenge could be built or *how* the Spartans could defeat Xerxes, for example. Basically these questions are 'why' questions asking for necessary conditions, not true 'how' questions, which ask for descriptions. But such questions also have an element of 'wherefore?' for they imply a request for the explanation of the actions of people.

As we saw in Chapter 7, there is more to explanation than the offering of causal accounts and the making of predictions and retrodictions. But, in so far as explanation does consist in giving us the power to predict and does induce some expectations then something in the D-N form is required. It is surely the case that we wish to predict, or at least have rational expectations of, birth rates, housing needs, food requirements and markets generally. We want to be able to retrodict as well: to know why high-rise flats were a failure, why teenage pregnancies have increased, why there was a move away from city centres, and so on. The fact that we have some success with prediction and retrodiction shows that there are some regularities on which we can at least in part rely and that the methodology imported from the natural sciences is at least to some extent effective. It may not be *as effective* in the human sciences as it is in the natural sciences, but it does make an important contribution to our search for explanations of human behaviour.

In the next chapter we shall consider the inadequacies of the D-N model in relation to understanding, and we shall also see that though it is easier to appreciate these inadequacies by appealing to instances from the human sciences, the natural sciences can also furnish us with examples.

12

Inadequacies of the D-N Model

In explaining a happening or a fact, one shows that it is not just some disconnected occurrence. A good explanation reveals the extent to which the happening to be explained is linked to some other more general phenomena. The more extensive and systematic the linkages (and thus the more general the phenomena to which the happening is linked), the better the explanation.[1]

Although a D-N explanation often does comprise laws supported by higher-level explanatory theories which enlarge conceptual understanding, its primary function is to show cause. This was the basis of Scheffler's criticism (see Chapter 6, p. 59). Therefore the D-N model can be inadequate and it can be inadequate both as an explanation in physical science and as an explanation for human sciences. By showing the nature of this inadequacy in more detail we shall justify criticism of D-N models and we shall also show more clearly the importance of the conceptual aspect of explanation, that aspect that offers answers to the question 'Wherefore?'.

We saw in Chapter 4 that a higher-level theory modifies concepts in lower-level laws it explains and so can change the significance of the words – in effect it introduces a new language. The D-N model of explanation (whether composed of laws supported by a higher-level theory or no) does not indicate that there may be modification of meaning inherent in the explanation and therefore it can be misleading. For example, as we saw in Chapter 5 Galileo's law of falling bodies gives a description of how bodies fall and also offers a causal explanation of their fall. That causal explanation rests on laws that assume that the velocity of fall is directly proportional to the time of fall so that the acceleration is constant. Now the laws can themselves be explained by appeal to a higher-level theory, the Newtonian theory of gravitational attraction. This further explanation is not demanded by the D-N model based on Galileo's laws, and the theory may be invoked without the full implications being stated: Thus the simple explanation would be:

Galileo's law states that all bodies accelerate uniformly as they fall at 32 ft/sec/sec.
This body falls for 2 secs.
Therefore its speed on impact is 64 ft/sec.

The law could be related to Newton's theory:

Newton's theory of gravitational attraction entails that all bodies are attracted to the earth by a constant gravitational force. This produces the acceleration observed in falling bodies.

Unless the Newtonian theory is fully articulated it is not apparent that the acceleration of falling bodies is not expected to remain constant. It happens that for bodies falling to the earth the increase in acceleration is small so that Galileo's laws give sufficiently accurate predictions in most circumstances, nevertheless the concept of acceleration due to gravity has been modified by Newton's theory. As we saw in Chapters 3 and 4 the language of the *explicans* differs from the language of the simple law and the language of the *explicandum*, and to understand the explanation properly it is necessary to understand the higher-level theory and to appreciate how the simple concept has been modified. The D-N explanation does not help us to do this because it is not formulated to show conceptual changes and changes in the meanings of words.

We can be misled by changes of meaning in explanations offered in the fields of the physical sciences but we are more likely to be misled by changes of meaning in explanations offered in the fields of the human sciences. This is because, as we saw in Chapter 5, the words used are less obviously technical (we think we know their meanings) and a certain amount of fluidity has (as we saw in Chapter 8) to be tolerated. Therefore it is not uncommon to find a 'slide' from one meaning to another as a person argues; and it is, unfortunately, common to find disputants unaware that their concepts differ though their words are the same. Consider an observed regularity, namely that suicide rates for Roman Catholics are lower than those for Protestants. It may be used as a simple causal law in a D-N explanation of there being a lower suicide rate in Spain than in Sweden:

> Suicide rates are lower for Roman Catholics than for Protestants.
> Spain has a much higher proportion of Roman Catholics than Sweden, which is largely Protestant.
> Therefore suicide rates are lower in Spain than in Sweden.

Such an explanation itself prompts further inquiry as to *why* Roman Catholics are less likely to commit suicide. One explanatory theory suggests that people lacking moral guidance and sympathy are more likely to commit suicide and that the Roman Catholic Church is organised in such a way that its priests are encouraged to offer positive moral guidance and sympathy. Using this explanation a new causal account can be given:

> Suicide is more common among those who lack moral guidance and sympathy.
> Roman Catholic priests are encouraged to offer moral guidance and sympathy more than are Protestant pastors.
> Therefore suicide rates are lower for Roman Catholics than for Protestants.

Just as Newton's theory modifies the Galilean concept of the acceleration of falling bodies, so this theory modifies the concept of suicide: it is no longer just self-destruction but self-destruction due to lack of moral guidance and sympathy. The explanatory theory may be further developed: it may be suggested that the moral guidance and sympathy given by Roman Catholic priests is part of Roman Catholic countries' religious and social policy to encourage contentment and preservation of the *status quo* and/or to encourage interdependence and recognition of mutual obligations. Suicide then acquires a social and religious significance because it implies that the suicide is alienated from society and perhaps that society itself is at fault.

In debate about suicide and suicide rates the different concepts may not be distinguished: the term 'suicide' may sometimes denote self-destruction, an act that is of concern to the individual and family but of no social concern; or it may be regarded as an act involving rejection of moral obligations, one to be condemned as showing lack of moral fibre; or it may be seen as an act of despair due to lack of help and therefore an indictment of society, a particular group or the social order. Debate will be confused unless there is prior discussion as to how 'suicide' is to be understood. It is only if concepts are clear that facts can be clear and that explanations can give us understanding and knowledge of the world and of ourselves.

In this text a realist view of knowledge is assumed – a view that our empirical theories are not entirely artificial and subjective constructs designed merely to correlate observations or to provide a convenient conceptual framework or paradigm. As we saw in Chapter 4, facts are theory-laden but in an important way they are independent of us: of our observations, our theories and our language. This realist view is dependent on a metaphysical belief that there is an objective reality and that theories are true in so far as they consist of propositions that correspond to that reality. Then, if a theory is accepted as true it will have educed facts; if it is rejected as false it will have educed fictions (see Chapters 2–4). Fictions, known to be fictions, do not change our judgements of the world though they can enrich our imagination and stimulate us with new ideas as to what might be and as to what may be possible. At any given time it is the theories currently accepted as true that are the bases of facts; they give significance to our words and so provide us with a meaningful language to describe the world. It is clear that words such as 'electron', 'gene', 'bourgeois', 'working-class' and 'monetarism' can only be significant if the educing theories are understood, and equally clear that they can be, and are, used without understanding; but much more familiar words like 'person', 'sun', 'electricity' and 'water' depend on educing theories for their significance.

Now it must be admitted that many theories deemed to be true in the past have subsequently been shown to be false; and that *all* theories are modified. A difficulty for realists is that when any theory is modified the meaning of its terms (the language of the theory) is also modified. For example, the word 'electron' as used in the language of Bohr's theory of the internal structure of atoms (early twentieth century) does not have the same meaning as the word 'electron' as used in the language of contemporary nuclear physics. But we do not have to conclude that the word now signifies an entirely different entity; we can say that the current scientific concept of electrons is a modification of Bohr's concept for Bohr's electron does have properties in common with, and other properties similar to, the contemporary electron; so physicists can relate Bohr's electron and his theory-language to their electron and their theory-language.[2]

As was indicated in Chapter 3, what appears to be a disagreement about facts may be a disagreement about theories; Quine gives an example:

> Picture two physicists discussing whether neutrinos have mass. Are they discussing the same objects? They agree that the physical theory which they initially share, the preneutrino theory, needs emendation in the light of an experimental result now confronting them. The one physicist is urging an emendation which involves positing a new category of particles, without mass. The other is urging an alternative emendation which involves positing a new category of particles with mass. The fact that both physicists use the word 'neutrino' is not significant. To discern two phases here, the first as agreement as to what the objects are (viz. neutrinos) and the second a disagreement as to how they are (massless or massive) is absurd.[3]

Later he says of more established terms (established because the educing theories have been accepted for some considerable time and are part of classical science):

> Even the identity historically introduced into mechanics by defining 'momentum' as 'mass times velocity' takes its place in the network of connections . . . if a physicist subsequently revises mechanics so that momentum fails to be proportional to velocity, the change will probably be seen as a change of theory and not peculiarly of meaning.[4]

I suggest that change of theory *is* change of meaning if the theory is being taken into account. As was explained in Chapter 3, it is always possible to describe in terms neutral between two (or more) higher-level theories by avoiding any modification that the competing theories might make to the simple concept. The meaning of 'momentum' might be unchanged for the non-scientist or schoolchild because she would be unaware of the new theory. The meaning of 'acceleration of falling bodies' is not changed by Newton's theory if that theory's implications are not understood. But if the explanatory theory involving a new concept of momentum is understood and accepted then it must modify the simple concept of momentum and the meaning of the word 'momentum' must be modified.

We can never be sure that our theories are true and that our explanations are correct, we must always be prepared to modify and even to reject them; but as realists we hold that our theories converge towards the truth and that our explanations give us a true account of events and are not merely providing formulae for reliable predictions and retrodictions or an expendable picture of the external world. This is to restate the metaphysical assumption that there is an objective reality which we can aspire to describe and (at least in part) to understand. Putnam expresses the realist view:

> realism depends on a way of understanding truth, not just a way of *defining* the word 'true'. The concept of truth is *not* philosophically neutral. . . . the distinction between total theory and term meaning is *useless*. . . . 'total theory', . . . means here just that: *total* theory, not just *logical* theory but total theory of knowledge; and this involves our theory of nature and of our interactions with nature.[5]

In the human sciences we have extra problems because, as we saw in Chapter 3 and in the account of suicide given above, our explanatory theories and the facts they educe are affected by our moral and aesthetic values. Our judgements about individuals and about human societies cannot be as value-free as are our judgements about the behaviour of inanimate objects and of other animals and plants. Our human world is changed when our values change – changed for *us*, because we change our interpretations and change our explanatory theories. Consider how changes in values have altered our views and therefore our accounts of marriage, abortion, illegitimacy, religion, sport and work in the last few decades. As indicated in Chapter 4, different theories produce different languages and since values affect theories we speak a different language with a change in values; this will be discussed at greater length in Chapter

14. An important inadequacy of the D-N model of explanation, related to the general inadequacy due to ignoring higher-level theory, is that there is no means of catering for the role of values. A D-N model aims to provide a causal account within a given context and does not indicate alternative contexts. Neither does it indicate that there may be an alternative or complementary physical explanation to a teleological explanation and *vice versa*. Putnam's example shows the importance of context:

> Willie Sutton (the famous bank robber) is supposed to have been asked 'Why do you rob banks?' His reply was 'That's where the money is.' Now . . . imagine (a) a robber asked the question; (b) a priest asked the question.[6]

Since the D-N model does not indicate that there *are* alternative accounts it cannot guide our choice of accounts or help us to find the appropriate one.

Another defect of the D-N model is that it does not do justice to our desire to understand particular events *as* particular events (albeit 'placed' in a larger scheme). This defect is particularly important in relation to many explanations sought in social science when we may be interested in finding an explanation for the behaviour of a particular person or group. We are interested in the particular *as a particular*, and not as an example of the general; the D-N model, depending on general laws, offers no help.

Lastly, many explanations in the social and human sciences cannot be satisfactory unless the beliefs and intentions of the agents featuring in the *explicandum* are known and understood. The D-N model does not indicate that there may be competition between accounts given by observers and accounts given by participants and hence cannot indicate how conflicting accounts are to be assessed. Neither does it guide us to understanding of accounts given by agents. It has been suggested that we need a special sort of understanding if we are to give a satisfactory explanation of human actions: we must be able to view actions as the agents themselves view them. This understanding is claimed to be given by the operation of *Verstehen*.

13

The Operation of Verstehen

In Chapter 1 it was stated that the operation of *Verstehen* is involved in any type of explanation that attributes conscious desires, expectations, beliefs and purposes to others. In this chapter that view will be developed and it will also be argued that the operation of *Verstehen* is involved in explanations of social customs and ceremonies, that is that it is necessary to know the significance of social behaviour to the participants and the explanations (if any) that they offer. There is this necessity even if, and perhaps especially if, it is thought that their accounts are incorrect. There is, however, no suggestion that the operation of *Verstehen* alone will lead to a satisfactory explanation; the suggestion is that *Verstehen* is an essential *part* of such explanations.

There is no general agreement on this point; Abel allows that the operation of *Verstehen* may be a help but implies that it is not essential. For him a satisfactory explanation must be based on external observation (leading to empirical laws and theories) and tests of the explanations offered. The operation of *Verstehen* may be of psychological help:

> the operation of *Verstehen* does two things: it relieves us of a sense of apprehension in connection with behaviour that is unfamiliar or unexpected and it is a source of 'hunches', which help us in the formulation of hypotheses.[1]

but

> The operation of *Verstehen* does not, however, add to our store of

knowledge, because it consists of the application of knowledge already validated by personal experience; nor does it serve as a means of verification. The probability of a connection can be ascertained only by means of objective, experimental and statistical tests.[2]

Schutz takes an intermediate position.[3] He points out that opponents of the operation of *Verstehen* are concerned with the need to collect objective empirical data and to develop theories relating to observable facts. They point to advances in such social sciences as economics where studies of curves of supply and demand, and equations of prices and costs, have been crucial, as opposed to vague conjectures about people's wants and values. In other words what *happens* (what can be observed and, hopefully, measured and quantified) is what is important; what people *think* is, at best, marginal. Schutz says that the empirical approach is based on idealisation, on ideal models of the social world, and he argues that a study of society based on attempts to understand the way its members understand it is to adopt an alternative approach. Thus in his view there are two alternative methodologies, two schemes of reference, the objective and the subjective, which are concerned with social groups and social persons respectively. A decision to adopt one scheme or the other confined the inquirer to one aspect of the social field. However, as Schutz points out, the two cannot be easily separated and it is difficult to comply with what has been called 'purity of method'. Moreover he thinks that the subjective methodology, the attempt to seek the actors' views, is essential if an explanation is to provide a basis for action:

> for a theory of action the subjective point of view must be retained in its fullest strength, in default of which such a theory loses its basic foundations, namely the reference to the social world of everyday life and experience. The safeguarding of the subjective point of view is the only but sufficient guarantee that the world of social reality will not be replaced by a fictional non-existing world constructed by the scientific observer.[4]

Even more vigorous support is given to the operation of *Verstehen* by Derden. He holds that it is logically necessary for the explanation of all purposive human behaviour:

On the basis of the schema presented, part of what is going on in empathetic understanding (when one 'puts oneself inside the head of the other') is that one is appropriating A's premises and seeing (logically) whether A's conclusion/decision follows. . . . *If I can infer A's conclusion from A's premises then I have empathized with A*. There is a logical seeing/understanding of the connection between A's decision/conclusion and A's reasons. . . .

This means that empathetic understanding can be used as a method, not for discovering a hypothesis, but rather as a law to predict. . . . In short, employing certain generalizations, empathetic understanding can be a method for discovering initial conditions [i.e. D-N minor premises]. . . . More important, empathetic understanding as a method of discovering initial conditions for predicting an agent's action is unique to the study of reasoning creatures. This is so because we as observers must deduce what the agent deduces, where the conclusion deduced is then needed to set up an initial condition. . . . One form of 'putting oneself inside the head of the other' is appropriating the other's premises and inferring his conclusion. The mysterious seeing is not mysterious at all; it is a matter of logic.[5]

It must be stressed that the operation of *Verstehen* does not require that we have the identical experiences of those whose actions we aspire to explain but that we must have enough experience and enough sympathetic imagination to be aware of the feelings (and emotions) of others and to be able to understand their reasons for their beliefs and their actions. For example, we need to have some experience of apathy and hopelessness and some imagination to project those experiences more deeply, to understand the apathy and hopelessness of those who no longer wish to live, or of those who have been unemployed for so long that they cannot make the effort to find or to do work. Without that understanding we cannot expect to offer a satisfactory explanation of suicide or of the behaviour of the unemployed.

It does seem eminently reasonable to maintain that explanations of human actions must involve appreciation of the significance of those actions for the actors themselves. However, as Trigg says, we have to ask whether such empathy is possible

Ethologists do not have to ask what is going on in the minds of monkeys washing potatoes; the monkeys' behaviour requires no further interpretation. Human action, however, can only be understood, it is suggested, in terms of human concepts. This inevitably raises the question of interpretation, since, if the nature of an action is determined by the consciousness of an agent, there is a difficult question how far an observer can become attuned to that consciousness.[6]

An objection raised by critics of the value of the operation of *Verstehen* is that explanations based on appeals to inner experiences of the agents (and understood by the inquirer(s) through the operation of *Verstehen*) are not subject to confirmation or to refutation by overt observation. Explanations offered in the physical sciences are based on empirical laws and theories which are, at the last, supported by observation; they can, at least in principle, be confirmed (or falsified) by all suitably placed and suitably qualified observers. By contrast inner experiences cannot *in principle* be observed by others, they are essentially private. This objection is related to the view that there cannot be a private language,[7] that is a language devised by an individual to describe both external objects and events *and* her inner experiences and used only for communication with herself: introspection and inner debate. Wittgenstein takes the view that a language is essentially public and that it *must* be related (directly or indirectly) to what can be publicly identified by ostensive definition (being pointed at). Wittgenstein argues that it is nonsensical to suggest that a language confined to one person could be significant, even for that one person. Thus talk about objects such as tables is significant because our agreement about the meaning of 'table' is based on ostensive definition; talk about inner experiences is significant because we can ostensively define the experience by 'pointing at' the characteristic behaviour that accompanies it. For example, any talk about pain is talk about observable pain behaviour; 'observable pain behaviour' is what 'pain' *means*. Wittgenstein does not deny that we have the inner experience of pain but he holds that anything we say (or write) about pain cannot refer to this inner experience; we have, he argues, no way of checking that the inner experiences are the same, always *pains*, save through the test of observable behaviour, that can be observed by others as well as by ourselves.

Let us imagine the following case. I want to keep a diary about the recurrence of a certain sensation. To this end I associate it with the sign 'S' and write this sign in a calendar for every day on which I have the sensation. – I will remark first of all that definition of the sign cannot be formulated. – But still I can give myself a kind of ostensive definition. – How? Can I point to the sensation? Not in the ordinary sense. But I speak or write the sign down, and at the same time I concentrate my attention on the sensation – and so, as it were, point to it inwardly. – But what is this ceremony for? for that is all it seems to be! A definition surely serves to establish the meaning of a sign. – Well, that is done precisely by the concentration of my attention; for in this way I impress on myself the connexion between the sign and the sensation. – But 'I impress it on myself' can only mean: this process brings it about that I remember the connexion *right* in the future. But in the present case I have no criterion of correctness. One would like to say: whatever is going to seem right to me is right. And that only means that here we can't talk about 'right'.[8]

The rebuttal of this argument has been given, in my view conclusively, by Ayer. Doubtless we do usually have observable behaviour 'checks' on inner experiences – always in the case of others and often for ourselves. Perhaps it is the case that we come to learn the significance of words such as 'pain' by relating them *at first* to observable behaviour. But, at some point, we are obliged to rely on our memory; we have to take something on trust. This applies as much to our recognition of external objects and events as to inner experiences. As Ayer says:

Let the object to which I am attempting to refer be as public as you please, let the word which I use for this purpose belong to some common language, my assurance that I am using the word correctly, that I am using it to refer to the 'right' object, must in the end rest on the testimony of my senses. It is through hearing what other people say, or through seeing what they write, or observing their movements, that I am enabled to conclude that their use of the word agrees with mine. But if without further ado I can recognize such noises or shapes or movements, why can I not also recognize a private sensation? It is all very well for Wittgenstein to say that writing down the sign 'E', at the same time as I attend to the sensation is an idle ceremony. How is it any more idle than writing down a sign, whether it be the conventionally correct sign or not, at the same time as I observe some 'public' object? [9]

and

But, it may be said, in the one case I can point to the object I am trying to name, I can give an ostensive definition of it; in the other I cannot. But what difference does this make? I can indeed extend my finger in the direction of a physical object, while I pronounce what I intend to be the object's name; and I cannot extend my finger in the direction of a private sensation. But how is this extending of my finger itself anything more than an idle ceremony? If it is to play its part in the giving of an ostensive definition, this gesture has to be endowed with meaning. But if I can endow such a gesture with meaning, I can endow a word with meaning, without the gesture.[10]

Ayer goes on to say that it is false to assume that for a person to attach a meaning to a sign (say for her inner experience) it is necessary that other people should be capable of understanding it too.

T. Nagel argues that inner experiences are neither *completely* private not *completely* subjective; he says that an account of any experience is an account of a *type* of experience:

I am not adverting here to the alleged privacy of experience to its possessor. The point of view in question is not one accessible only to a single individual. Rather it is a *type*. It is often possible to take up a point of view other than one's own, so the comprehension of the facts is not limited to one's own case. There is a sense in which phenomenological facts are perfectly objective: one person can know or say of another what the quality of the other's experience is. They are subjective, however, in the sense that even this objective ascription of experience is possible only for someone sufficiently similar to the object of ascription to be able to adopt his point of view – to understand the ascription in the first person as well as in the third, so to speak.[11]

Later Nagel develops the view that it may become possible to arrive at new and more objective concepts of experiences; but this possibility, he says, does not detract from the importance of the operation of *Verstehen*, the importance of imagining how others feel, in explaining human behaviour at present:

> At present we are completely unequipped to think about the subjective character of experience without relying on the imagination – without taking up the point of view of the experiential subject. This should be regarded as a challenge to form new concepts and devise a new method – an objective phenomenology not dependent on empathy or the imagination.[12]

Whether any attempt to meet the challenge would be successful remains an open question but it is certainly arguable that it must require more than observation of external behaviour and of internal physiological events to understand the role of another's conscious awareness (as opposed to giving an account of physical and chemical events in another's brain). Sufficiently detailed physical knowledge of brain events is not available today but if (and when) there is detailed correlation between conscious experience and brain activity it may still be the case that the latter does not explain the former and so cannot explain all aspects of human action. It may be that attempts to explain human behaviour in the same way as the behaviour of inanimate objects are based on a fundamental misconception. Such explanations would provide a *kind* of understanding that might be useful in certain contexts, for example in medical diagnosis and treatment, though at least some doctors may wish to appeal to more than physical (neurological) events. Writing of the accommodation to nervous disease, Sacks says:

> Accommodation, though universal . . . is the least discussed, the least understood, the most mysterious of phenomena – the final transcendence of 'I' over 'It', person over mechanism. What starts out, what started out, as a modification of mechanism, finally calls forth the highest powers of the self – and shows us the role of the transcendental in medicine, and the form of a medicine which transcends medication.
> This, finally, is increasingly acknowledged, even by the most prosaic and hard-boiled titrators: what is defeat in the realm of the mechanical, the problematic, our inability to solve an algebraic x,

may bring us to the reality of a transcendental x – the need to feel this, and woo it, and give it its due.[13]

and writing of a patient who could remember nothing for longer than a couple of minutes:

> I had wondered when I first met him, if he were not condemned to be a sort of 'Humean' froth,[14] a meaningless fluttering on the surface of life, and whether there was any way of transcending the incoherence of his humean disease. Empirical science told me there was not – but empirical science, empiricism, takes no account of the soul, no account of what constitutes and determines personal being.[15]

Our discussion in Chapter 10 (pp. 89–90) suggests that explanations of animal behaviour may also be inadequate without appeals to intentions but there can be no doubt that such appeals are essential for explanations of human actions. Behaviourists, those who regard appeals to consciousness and to intentions as misleading, are in effect claiming that there is no distinction between actions and happenings when we seek explanations. Both must be assessed as observable physical events in order to be subjects for scientific study; what goes on in the mind cannot be observed and is therefore irrelevant. But perhaps they are ruling out an essential constituent of human behaviour (and possibly animal behaviour); what human beings *think* may be an intrinisic part of what they *do*.

> One of the issues, indeed, is just how important human consciousness is. Does it constitute human social activity by the meaning it bestows on it, or is it irrelevant? One of the problems of dismissing consciousness and its products as perhaps an interesting but irrelevant appendage to human activity is that this must be done through arguments emanating from human consciousness. The extreme materialist who says that ideas are unimportant is, in fact, in the grip of an idea.[16]

and

> Human beings do not just *behave*. They *act*, and their actions occur with an understanding of their significance in a wider social

context. Human actions are endowed with meaning, and it may appear that they cannot be properly understood unless that meaning is grasped.[17]

If this is so the operation of *Verstehen* is essential.

The operation of *Verstehen* also involves developing practical knowledge, analogous to the knowledge we have that enables us to walk or to play the piano. Psychological knowledge also embodies practical knowledge, a skill in assessing as opposed to a skill in performing. Thus *use* of a psychological explanation and description involves more than appealing to a psychological theory:

> It may be perfectly clear to everyone in a given situation that Jones is jealous of Smith's reputation. But one couldn't give *anything like* a 'scientific proof' that Jones is jealous of Smith's reputation. . . . one can't 'verify' *Jones is jealous* in isolation: one would have to verify a *huge* 'psychological theory' which covered all the 'special circumstances'. And this, of course, is implicit in our knowledge of people and our *ability* to use psychological descriptions – not something we can state explicitly. Again, it is a feature of 'scientific' knowledge (at least if we take physics as a paradigm) that we *use measuring instruments that we understand*. Our theory applies to our measuring instruments, and to their interactions with what they are used to measure, not just to the objects we measure. It is a feature of *practical* knowledge that we often have to use *ourselves* (or other people) as measuring instruments – and we do *not* have an explicit theory of these instruments.[18]

The human sciences are concerned with the actions of human beings and their investigations are themselves human actions. The scientist has to assume that certain theories and certain propositions about people's thoughts and feelings are true (that they express facts, see Chapter 4) and an important criterion for judging the truth of such propositions is founded on appeal to our own experiences and the belief that others can and do have similar experiences. We cannot 'check' this belief against some objective standard, that is against what are generally held to be more public data obtained by overt observation, and measurement involving comparison with an established scale. The operation *Verstehen* supplements, and indeed provides a necessary supplement to, explanations in the human sciences. It can give us something much better than a possible explanation, particularly when it is linked to other aspects of explanation.

> empathy gives less than 'Knowledge' . . . but it gives more than a mere logical or physical possibility. It gives *plausibility* – it is a source of prior probability in many judgements about people . . . I am arguing that knowledge *depends on* a good deal of right opinion. *Verstehen* is a source of prior probability; the 'hypothesis' to which we assign a significantly high prior probability on the basis of *Verstehen* (empathy) must indeed by 'checked', but *even this checking is ultimately intuitive.* The alternative picture of social science – as based on 'general laws of history, sociology and psychology', . . . and applied to data as in physics is false . . . and utopian as a vision of future social.
>
> The purpose of Nagel[19] *et al.* was to rule out what they viewed as 'obscurantism' and 'metaphysics' in the social sciences. Worthy as these aims are, to pursue them by misrepresenting what actually goes on is to promulgate an ideology, in the pejorative sense of 'ideology'; not to promote clarity of method.[20]

At the end of the day we have to trust our own judgement – 'this checking is ultimately intuitive' – and, as Ayer pointed out, any appeal to observation (as in the physical sciences) is, at the last, dependent on our reliance on our own intuitive judgement. Putnam holds that the social sciences cannot resemble the physical sciences since we cannot be completely objective when we study ourselves; this is not something that should, necessarily, be regretted:

> Should we *regret* the fact that the social sciences cannot realistically hope to resemble physical science? To ask this is to ask if we should regret the fact that we cannot understand ourselves and each other as the physicist understands the harmonic oscillator. . . . If we are doomed to have neither a computer's-eye view nor a God's-eye view of ourselves and each other, is that such a terrible fate? We are men and women; and men and women we may be lucky enough to remain. Let us try to preserve our humanity by, among other things, taking a human view of ourselves and our self-knowledge.[21]

Because we cannot view ourselves as inanimate objects it does not follow that human sciences can make no claim to objectivity. As we have seen we can establish some objective laws and quasi-laws applicable to human actions; the subjective aspect of explanations rests on the appeal to the imaginative exercise of empathy. But many human actions take place in a social context and to understand them more than imagination is needed, we need to be informed of the social and historical background. For example, the significance of firing guns at an IRA funeral is very different from the significance of firing guns on the occasions of the birth of a royal prince. Unless the social and cultural concepts are fully understood it is not possible to provide an explanation for that explanation must involve those very concepts. The greater the difference between the culture of the observed and the observers the more difficult it is for the latter to explain the behaviour of the former, and yet the more important it is to see over cultural barriers and to carry out the operation of *Verstehen*. Thus another society may differ from British society in respect of industrialisation, as does Afghanistan, in respect of wealth, as does India, or in respect of penal code, as does Saudi Arabia, in respect of social values, as did Hitler's Germany or present-day USSR, or in respect of religious influence, as does Spain. Any explanation of social customs, ceremonies and behaviour generally must have allowed for attitudes very different from those of an ordinary Briton; alien attitudes may seem strange, even repugnant, but they have to be understood from the point of view of the people concerned and they certainly cannot be ignored. It does not follow, as we saw in Chapter 2, that alien attitudes must be adopted, it does follow that they need to be known and understood.

Within our own society background knowledge is assumed but explanations can be needed for children and for strangers: explanations of queuing for a bus, of tipping in a restaurant, of drinking the loyal toast, of placing a bet, and so on all depend on knowledge of social customs and also on theories and evaluations of social behaviour. Such explanations will appeal not only to the general custom but also to the view taken (by society) of the morality of taking one's turn, of rewarding service, of the institution of Royalty, of the social toleration of alcohol and of the facilities and opportunities for placing bets. Different observers with the same knowledge but different values will offer different, possibly complementary, possibly incompatible, explanations and all of these might differ from explanations given by the participants (see also Chapter 1).

Suppose, for example, we are asked to explain why so many people are Roman Catholics[22] – we have to understand what their faith means to the believers and also how *they* think they came by their beliefs – their reasons for their beliefs. This will involve us in learning to understand their concepts of religion, of God and of the Church. Different inquirers, Marxists, Calvinists, Anglicans, agnostics and atheists, will have their own separate, possibly very different concepts and will have varying degrees of difficulty in carrying out the operation of *Verstehen*. In seeking understanding of Roman Catholic beliefs they may come to reassess their own beliefs and attitudes and so the operation of *Verstehen* can be doubly advantageous in the promotion of understanding.

In this chapter it has been shown that the operation of *Verstehen* is required for adequate explanations of human actions. It follows that the human (including social) sciences must be less objective and more complex than the physical sciences though in both fields explanations are based on appeals to regularities (laws and quasi-laws) and the D-N model is applicable. A further complication in the study of human behaviour is the inevitable intrusion of values such that fact and value cannot be clearly distinguished. This was referred to in Chapter 1, p. 8, and in Chapter 3, p. 33 and we shall now consider the matter more deeply.

14

Values and Explanation

We have seen that all knowledge in all the sciences, that is all empirical knowledge, depends on theories, and that, especially in the human and social sciences, the theories will reflect and embody values, particularly moral values. What we regard as right and wrong greatly influences our assessment of any explanatory theory involved in the interpretation and explanation of human behaviour. Thus those who believe that eating animals is wrong and who are vegetarians will judge the economics of farming and the eating habits of various social groups differently from non-vegetarians; they are very likely to offer different accounts, and therefore different explanations of, say, famines and food shortages and the organisation of marketing farm produce from those offered by meat-eaters. Likewise Marxists[1] will give different explanations of, say, inner-city riots from those offered by capitalists, and the explanation of the popularity of beauty competitions suggested by feminists will differ from that offered by the promoters. Our values provide us with at least some of the criteria for accepting (and for rejecting) explanatory theories.

The sociologist Max Weber (1864–1920) distinguished two types of knowledge: existential knowledge, knowledge of what *is*, and normative knowledge, knowledge of what *ought to be*. The former was concerned with means and was the proper field for social science; the latter was concerned with ends and related to advice about social policies. Weber appreciated that social scientists would almost certainly be concerned with policies in social matters and that social problems would involve debate about ends as well as means. He said:

the more 'general' the problem involved, . . . the broader is its cultural *significance*, the less subject it is to a single unambiguous answer on the basis of the data of empirical sciences and the greater the role played by value-ideas (*Wertideen*) and the ultimate highest personal axioms of belief.[2]

But he was anxious to preserve the distinction just because it was so easy unwittingly to slide from 'is' to 'ought' (and *vice versa*). He did not think that values were unimportant but he wanted them to be introduced overtly; they were needed but were not to be confused with facts:

> The constant confusion of the scientific discussion of facts and their evaluation is still one of the most widespread and also one of the most damaging traits of work on our field. The foregoing arguments are directed against this confusion, and not against the clear-cut introduction of one's own ideals into the discussion. An *attitude of moral indifference* has no connection with *scientific* 'objectivity'.[3]

Weber was writing in 1949; since then it has come to be appreciated that fact and value cannot be clearly separated. Kaplan, writing in 1964, supported Weber's view that values should not be concealed and were part of the sociologist's field, 'Freedom from bias means having an open mind, not an empty one',[4] but he argued that values were inevitably involved in the choice of problems and inextricably involved with facts because 'values enter into the determination of what constitutes a fact'.[5]

Moral values become embodied in beliefs about what *ought* and *ought not* to be done and they are also an influence on what we think *is* and *is not*. So it can be very difficult to make a clear distinction between beliefs (ideologies) and facts. Ryan alludes to Marx's view that a person's social background affects her whole outlook so that her interpretations and explanations of social customs and behaviour are a result of something much deeper than mere bias; they are based on her whole way of life

> It is not simply that the social position of a writer or a scientist will prejudice him in favour of some beliefs and against others. It is also the assertion that a person's whole way of life is bound up with the

way he thinks, in such a way that thought and action together form an integral whole.[6]

Ryan points out that the view that the way we think is bound up with our way of life now goes 'far beyond the ranks of Marxist sociologists'. Others would agree with Marx that a dominant social class hires intellectuals, priests, artists and writers to promote its own social view, its own view of the truth, as *the* social view, as *the* truth. This is not imposed by force but pervades the social atmosphere as unexpressed but unquestioned assumptions: the nineteenth-century proletariat in Europe, immigrants to America in the early twentieth century, and perhaps women in Britain in the later twentieth century suffered or suffer from the atmosphere of value assumptions that is all the more effective for *not* being imposed by force. Ryan himself does not think it impossible to cut free of the prejudices of the social environment but he does admit that there is a problem for those who seek value-free or value-neutral accounts. This problem is highlighted by the ambiguities of language which is inevitably partly evaluative. Words and concepts such as 'democracy', 'bourgeois', 'inflation', 'unemployment', are value-laden.

The evaluative and cognitive elements of theories of human behaviour are intrinsically connected and therefore conceptually connected:

A change in moral emphasis – in priority among values – unavoidably demands a new way of sorting out facts and vice versa.

For instance, it has been said that, before Marx, poverty was just a personal misfortune, whereas after him, it was part of the oppression of the working classes. If we try to sort out the factual from the value elements in this change, we shall not find that they are logically isolated. Nobody could have made one change without the other. Marx – or somebody like him – was needed to do both.[7]

Even a description, a presentation of the facts of a situation, must be value-laden and an explanation of what is described is bound to reflect the embodied values. The importance of values is stressed by Thomas:

Could two social scientists, one of whom describes Britain as a liberal democracy and the other of whom describes it as a capitalist state, put forward the same explanations and the same evaluations of British political processes? Disputes about description are basically disputes about classification, about what objects in the world the theory recognises.[8]

and

> A social scientific theory necessarily includes descriptions of the phenomena in question. These descriptions cannot be detached from the theorist's explanations or evaluations, so that the theorist might take one line in his descriptions and another either in his explanations or evaluations. It is implausible to suggest that a theorist might describe the police in Britain as 'the main element in the repressive state apparatus' and then theorise about them as 'the enforcers of law and order'.[9]

It is much more difficult to separate the *explicandum* and the facts it presents from the *explicans*, and the effect it has on the facts it explains, when we are considering human sciences as opposed to physical sciences (see also Chapter 3). This is because in the human sciences the moral evaluative element is generally relevant and sometimes very important, and this affects the lower-level theory pertaining to the *explicandum* as well as higher-level theories of the *explicans*. Harding argues that this is a distinctive feature of human sciences:

> What marks phenomena as social actions rather than as biological, chemical, or astronomical events is that they strive to reach normative standards. They are explicitly or implicitly not only goal-directed, but directed towards normative goals. Consequently ... one cannot even identify *which* phenomena are social phenomena nor what kind of social phenomena they are without grasping the normative[10] concepts involved for the persons or culture observed. Here identifying value judgements are required.[11]

Hence, without knowledge and understanding of the actors' values the inquirer cannot provide a satisfactory explanation of their behaviour:

> if the observer's home society or the one observed *lacks* a range of

social concepts, it is probable that his description of the social regularities in the alien culture will be incomplete and/or biased by the gap in, or superfluous presence of, these values in his home society. (By 'home social values' I mean the inquirer's moral/political values. By 'alien social values' or 'alien culture' I mean simply values or culture not the inquirer's own. Alien values and culture sometimes exist within one's own political community, of course: the rich and the poor within a community might have different values and culture or might share the same values and culture.)[12]

The citizens of any state are members of subgroups with different cultures and interests and these can be supported by different values such that the members of one subgroup would regard the members of another subgroup as having alien values. In the nineteenth century Disraeli argued that Britain was two nations, each with a different culture:

'. . . our queen reigns over that greatest nation that ever existed.'
'Which nation?' . . . 'for she reigns over two.' . . .
'Yes,' . . . 'Two nations; between whom there is no intercourse and no sympathy; who are as ignorant of each other's habits, thoughts, and feelings, as if they were dwellers in different zones, or inhabitants of different planets; who are formed by a different breeding, are fed by a different food, are ordered by different manners, and are not governed by the same laws.'
'You speak of –' . . .
'The Rich and the Poor.'[13]

Disraeli implied that the cultural and social division *arose* from the difference in money, and it probably did. In Britain today the position is more complex since there is not the same vast difference between the prosperous and the needy and some important values are common to most people: attitudes to violent crime, to child abuse, to freedom of speech and the rule of law are similar. As was implied in the last chapter, we also share values that give us the same attitudes to certain conventions of behaviour such as queuing and, even if we do not all have the same attitude to, say, tipping or to the Royal Family, we readily understand the attitudes of those with whom we differ. However, values reflecting political attitudes do vary, as do our attitudes (as always based on values) to punishment, government control, defence, education and treatment of animals, to give but some examples. As it happens, those favouring any one political party do not all have the same values or take the same attitude on other issues; there is overlap. Partly because of this overlap, partly because some fundamental values are shared and partly because of shared conventions we remain *relatively* value-stable. However, even in a relatively stable society such as ours there are particular groups: trade unionists, merchant bankers and soldiers (whose actions we may seek to explain) with values that *in the fields with which they are concerned*, are alien to those outside the group. Moreover the values that they overtly support may differ from those that they actually support; and they may be unaware, or only half-aware, that this is so. Harding points out that in seeking explanations it is not sufficient to appreciate the actors' expressed values because sometimes they may themselves not appreciate their values, their beliefs and their intentions.[14]

> But an adequate explanation of an act often requires that the inquirer go beyond merely identifying and classifying the act as the actor would. For there are many cases in which an agent does not fully understand the true causes or consequences of his acts. And many acts which seem perfectly rational to the agent, appear irrational to an observer. If the social inquirer asked only the agent for an identification and classification of the act, he would often end up with an incomplete or irrational account. Now the inquirer must critically examine from the perspective of his own social value-system – that is, from the perspective of his own moral/political intuitions – the function that the act and the agent's kind of account of it have in the alien society. Otherwise, suspending such a critical examination, the inquirer biases his account in favour of the agent's value-system and produces just the kind of too-tolerant, particularistic account which scientific inquiry is designed to avoid.[15]

As we see, Harding does not suggest that the values of the inquirers are irrelevant – a value-neutralist account, that forbids appeal to the values of the inquirers cannot, in her view, be satisfactory. Alien values have to be known and understood but objectivity, in so far as it is possible, comes from taking the values of both parties into account:

We all know of notorious cases where the unexamined values of observers led them to mis-identify and mis-explain alien phenomena, characteristically where there was either great 'cultural distance' or a real conflict in interests between observer and observed. Missionary anthropologists misperceived much behaviour of non-Christian groups because they uncritically applied their own normative standards to societies with very different normative standards. Social scientists coming from societies with racist, sexist or classicist belief systems sometimes have misunderstood the beliefs and behaviour of persons in their own societies as well as of people in societies with very different attitudes toward race, sex or class. And inquirers from scientific cultures have had trouble accurately describing the beliefs and practices of traditional societies. While these sorts of cases lead the value-neutralist to argue that it is crucial that the inquirer *suspend* his values when he conducts inquiry, that the inquirer's values be banned from inquiry, the argument here is that from a scientific point of view such an approach is not only ineffectual but also incoherent.

Consider that if temperature is a variable which affects the volume of solids and we want to measure the volume of a solid, we do not try to 'suspend' temperature variations by ignoring them or not mentioning them, but instead to identify, evaluate the effect of, and control the variations. Refusing to examine the effect of a variable on a phenomenon does not contribute to increasing reliability of the results of inquiry; it decreases the reliability to the extent that we cannot predict how the variable effects [sic] the phenomenon. Similarly, if the values of the investigator help to shape the descriptions of social nature, then the values of the investigator should be identified with their effects.[16]

Harding suggests that the values of inquirers help to produce the phenomena they observe – however discreet anthropologists and sociologists are, it is likely that they influence the behaviour of those whom they observe. Thus we must take into account the values of both the observed and the observers. Of course this is not to affirm that moral values are the only factors relevant to explanations in the human and social sciences but it is to support what was stated at the beginning of this chapter, namely that moral values are part of the criteria that guide our choice of explanatory theory and therefore our choice of what and how we explain. The world which we both make and 'get at' through our theories has to be consistent; our many theories make what we hope will be a coherent system of empirical knowledge in which moral (and aesthetic) values will have their place.

Therefore it does not follow that there is no hope of objectivity and that, as some radical social scientists have affirmed, objectivity is not even a proper aspiration for social scientists. That attitude is attacked by Nisbet;[17] he concedes that inquiry in *any* science will never be completely free from personal predilections but, he says, this does not mean that there can be no distinction between inquiries that presuppose and use such predilections and those that strive to keep them to a minimum and also strive to avoid covert introductions of values.

It might, of course, be maintained that the connection between values and facts, and values and explanatory theories was contingent so that it ought to be possible, at least in principle, to give descriptions of human actions in value-neutral terms. The arguments in this chapter should, however, show that the connection is deeper, a conceptual if not a logical relationship. We must therefore be prepared to at least entertain the notion that there may be different, perhaps complementary, perhaps incompatible descriptions and explanations arising from different values. They could, in a logical sense, be equally valid:

> Where all the points of view on a physical phenomenon should be ultimately compatible at any given historical moment, a single coherent account of social phenomena is probably neither possible nor desirable. Interests are not just different; they are essentially conflicting in a way perceptions from different spatial locations are not.[18]

As Putnam said,[19] it is not a matter for regret that we are human and it is because we are human that we make moral judgements and take moral attitudes about human behaviour. We do not take a *moral* interest in animate objects, hydrocarbons, quasars, metals and so on, nor in the behaviour of plants and animals[20] but we not only do, we must, morally evaluate ourselves.

15

The Ultimate Explanation?

The account of explanation offered in this book portrays explanations as attempts to increase understanding by helping to co-ordinate our conceptual schemes about the nature of the world – the world of inanimate objects, and material substances, of animate beings, including ourselves, and of observable events. In addition it has been shown that explanations can enable us to predict and perhaps to control behaviour and events. It has been suggested that there are three aspects of explanation:

(1) The descriptive . . . answering the question 'How?'
(2) The causal / predictive . . . answering the questions 'Why?' and 'What?'
(3) The conceptual . . . answering the question 'Wherefore?'

It has been shown that all facts, and therefore all empirical descriptions, are theory-laden, so that any fact is taken to be a fact only because certain theories are understood and accepted as true. Even simple descriptions of *how* things are, depend on theory. We start our descriptions at the level of common sense and we appeal to basic theories that arise spontaneously and which give us our concepts of everyday objects, of the people we meet and of the society in which we live. These fundamental theories provide bases for our common-sense interpretation of the world; they give us hard data. Higher-level theories, devised to account for and to explain our common-sense world, our hard data of basic facts, may be just as much relied on as the fundamental theories but they are more readily acknowledged to be theories. As was pointed out in Chapter 3, it may be difficult for us to accept that apparently purely factual sentences such as 'This is a table' or 'That is the sun' embody and depend on theories about physical objects and their nature, though undoubtedly they do. But it is not so difficult to accept that factual sentences such as 'The earth travels round the sun' or 'Halley's comet appears once every seventy-six years' embody and depend on theories of the positions and movements of the heavenly bodies. Higher-level theories may be well known and may be very firmly established, but because they are learned and do not arise spontaneously they are more readily perceived to be theories; facts depending on them are more readily perceived as being theory-laden. But should an established theory come to be rejected (as was Ptolemy's theory of the sun-centred cosmos) then facts embodying them are facts no longer and any influence they had on lower-level facts supported by fundamental theories will disappear. This applies to theories of human behaviour as much as to theories of inanimate objects, to the human sciences as well as to the physical sciences.

We have seen that there are special difficulties associated with finding explanations of human behaviour, be it the behaviour of individuals or of societies though it does not follow that there is no place for a D-N type explanation in these fields. The D-N model can account for the descriptive and causal/predictive aspects of explanation; it serves well in the physical sciences and since intentions, beliefs and (indirectly) reasons may be shown to be causes it can be applied to explanations in the human/social sciences because, as we saw in Chapter 12, it fails to allow for the conceptual aspect of explanation and for the influence of higher-level theories. It ignores differences in language between the theory educing the facts to be explained (the language of the *explicandum*), and the theory educing the explaining facts (the language of the *explicans*). It is here that the special difficulties for the human sciences show: the data are more complex, the laws are less definite (usually quasi-laws) and there is the problem of reflexivity, the explanatory theory altering the *explicandum*.

Another difficulty which makes explanations in the human sciences different from those of the physical sciences is that it is necessary to undertake the operation of *Verstehen*, and in particular to allow for the value element that is inevitably present in descriptions and explanations of human behaviour, let alone in its assessment; there is a subjective aspect to the explanation of human behaviour that cannot be completely eliminated. But, having acknowledged this, we can also claim that explanations of ourselves need not be completely subjective and that with care the effects of our own prejudices and in-built assumptions can be controlled and limited.

Is it then possible to arrive at an ultimate explanation or a set of ultimate explanations? These would be accounts of the world which were themselves so complete that no further questions would need to be asked, understanding would be complete. Such an explanation or explanations would be a description, or descriptions of all entities and events that have occurred, are occurring and will occur – the 'How', 'Why' and 'What' would be subsumed into an all-embracing answer to the question 'Wherefore?' The language of such an ultimate *explicans* (or ultimate *explicantia*) would be the language of an ultimate theory (or theories). It is suggested here that it is unlikely that we can ever arrive at an ultimate set of explanations and even more unlikely that there is just one all-embracing ultimate explanation. An ultimate explanation, a single satisfactory explanation of *all* events, presupposes that all occurrences (physical events, behaviour of inanimate and animate entities, behaviour of individuals and of societies) are of the same kind. A set of ultimate explanations would show them to be inter-dependent and though this seems less bizarre it still seems far-fetched to suppose that no further questions could arise.

But even though the ultimate explanation or explanations are unlikely to be attainable we can still strive to find wider and more complete accounts of the world. We may be sidetracked from time to time but we can hope that, on balance, we are discovering more and more and understanding more and more. Our ideal is to arrive at true descriptions and a more comprehensive ordering of our experiences but to attain an *ultimate* description, one that satisfied all questions and therefore ended all questioning, would be stultifying. It is a paradox of human nature that our ideal is not only unattainable but also one that we wish to remain as unattainable for without curiosity and a sense of wonder, without speculation and imagination we would be diminished as human beings and the world would be prosaic rather than challenging. There would be no place for curiosity and no need to explore; there would be nothing to discover and nothing new to try to understand.

Glossary

antecedent: the former part of a hypothetical statement, e.g. 'p' in 'If p then q'. See also 'consequent' below.

cause – efficient: one of the four Aristotelian causes (material, formal, efficient and final). The efficient cause is the source of movement and/or activity, an engine or muscles, but *not* an intention; efficient causes are physical causes.

– final: (see above); the final cause is the reason or purpose in bringing an event about; it is *non*-physical and could be an intention.

condition – necessary: a condition which must obtain if an event is to occur or a statement is to be true; for example air being present is a necessary condition for human life.

– sufficient: a condition which will bring about an event if it obtains; for example decapitation is a sufficient condition for death.

consequent: the latter part of a hypothetical statement, e.g. 'q' in 'If p then q'. See also 'antecedent' above.

contingent: something which just happens to be the case, not a matter of logical necessity. A contingent attribute is not a defining attribute: thus possession of hair is a contingent attribute of mammals. A contingent truth is a statement that happens to be true, e.g. 'The earth is round'.

deductive-nomological (D-N): a form of explanation given as a logical argument. Certain laws (or one law) plus particular condition(s) stand as premises, and the conclusion entailed is the explanation.

empirical: based on observation (sense perception).

entailment: a logically necessary sequence of statements; the conclusion of a valid argument is entailed by its premises.

epistemic: of knowledge and of the grounds for knowledge.

epistemological: of the theory of the grounds for knowledge and the methods for obtaining knowledge.

explicans: that which explains.

factually necessary: of an event that *as a matter of fact*, does invariably occur in certain given circumstances, but which is not *logically* inevitable; compare below with 'logical necessity'.

helium: an element; one of the inert gases in the atmosphere but first detected in the atmosphere of the sun by spectral analysis.

logical necessity: of a statement that follows logically (i.e. is entailed by) accepted premises. It is to be contrasted with factual necessity, see above.

logical positivists: philosophers who believe that, apart from logic and mathematics, the only meaningful statements are those that can (in principle) be verified, directly or indirectly, by observation. For them metaphysical statements are meaningless.

metaphysics: going beyond physics, and all empirical knowledge; the concepts of space and time that provide the framework for physics and the other sciences.

micro-organisms: living creatures too small to be seen without the aid of a microscope.

natural necessity: see 'factually necessary' above.

ontological: of existence and of the manner of existence.

organic: pertaining to living organisms; also sometimes used to refer to carbon compounds (including those synthesised in laboratories).

ostensive definition: definition by direct indication of the entity concerned; e.g. 'red' would be 'defined' by pointing to red objects; a horse would be ostensively defined by pointing to a horse.

pH: see Chapter 5, p. 45.

premiss: a statement in a deductive argument that provides grounds for the conclusion.

retina: the light-sensitive nerves forming a screen of tissue at the back of the eyeball.

sense perception: the activity of the senses: seeing, hearing, touching, smelling, tasting, also kinaesthetic sensing.

spectrum: the band of colours into which a beam of white light can be dispersed by a prism. This is the visible spectrum (the colours of the rainbow), but radiation beyond the red and the violet (infra-red and ultra-violet) can also be dispersed to make the complete spectrum. Each element absorbs different parts of the spectrum when heated, producing characteristic dark lines; from the pattern produced an unknown element can be identified.

telos: purpose.

teleological: see Chapter 10.

truth – coherence theory: a statement is true if it is consistent with other statements accepted as true.

– correspondence theory: a statement is true if it *corresponds* to what is the case.

Notes

1 EMPIRICAL EXPLANATIONS

1. G. H. von Wright, *Explanation and Understanding* (London: Routledge & Kegan Paul, 1964), p. 6.
2. Explanations of some kinds of animal behaviour may imply excuse but we shall not discuss these here.
3. False explanations are not always so helpful; for example in early times it was thought that mad people were possessed by devils, demonic possession *explained* madness. The 'cure' was beating – to drive out the devils.
4. There were philosophers in Ancient Greece, Aristarchus for example, who proposed a heliocentric theory of the universe and who suggested that the earth rotated. It was a serious proposal, but there was little support for it because it was incompatible with other physical laws that were accepted at the time.
5. There are important differences between our present cosmology and that of Copernicus but the relevant similarity is the central position of the sun in the solar system.
6. See Chapter 2, pp. 22 *et seq.*
7. See D. C. Dennet, 'A Cure for the Common Code?', in *Readings in the Philosophy of Psychology*, vol. II, N. Block (ed.) (London: Methuen, 1981), p. 69: 'It is a mistake to make laws of the unreduced sciences exceptionless.' He is glossing another author, and agrees with him. See also A. Flew in *A Rational Animal* (Oxford: Clarendon Press, 1978), Chapters 3 and 7–9 and H. Putnam in Block, cited above, vol. I, Chapter 7 on the contingent necessity of physical laws.
8. For further discussion see Chapter 10, pp. 89 *et seq.*
9. R. Trigg, *Understanding Social Science* (Oxford: Basil Blackwell, 1985), p. 43.
10. In this text the term 'physical science' refers to sciences concerned with explanations that rest solely on physical laws and theories. It therefore includes subjects such as genetics and zoology and might be replaced by 'natural science'. However, explanations in certain natural sciences, such as psychology and branches of medicine, do not (at least according to some distinguished practitioners) rest solely on physical laws. Hence I

prefer to use the term 'physical science' and so to emphasise the particular distinction I wish to make between human (including social) sciences and those sciences that do not appeal to actions.

2   FACTS AND CONCEPTS

1. *The Times*, 15 September, 1982.
2. *The Times*, 6 January, 1983.
3. This is not to assert that the world is screened by a veil of sense data; to admit that there is inference is not to deny direct access. See A. J. Ayer, *The Central Questions of Philosophy* (London: Weidenfeld and Nicolson, 1973), Chapter IV.
4. In addition we have sensations arising from the physical events in our own bodies, such as kinaesthetic sensations, that we also interpret.
5. It is clear that many animals possess concepts, for they show by their behaviour that they recognise particular things as things of a certain kind, for example the dog recognises a bone as a desirable object, something to acquire. For the dog a bone may be a different object from the bone we see, but this is immaterial to the possession of a working concept.
6. Ayer, *Central Questions of Philosophy*, p. 207.
7. Nevertheless a common language does not guarantee mutual understanding for, as has been implied above, the same words in a common language can refer to different concepts and so have different meanings. Speakers can have different sublanguages within the common native language.
8. W. V. O. Quine, *Word and Object* (Cambridge, Mass.: MIT Press, 1964), p. 8.
9. It is possible that there are societies using different concepts in such a way that we might think we understood their language, and had translated successfully, when we had not; Quine discusses this possibility in Chapter II of *Word and Object*. However, we shall not allow for that possibility here; we shall assume that if speakers of different languages behave as though they understood each other then there has been successful translation.
10. D. J. O'Connor, *The Correspondence Theory of Truth* (London: Hutchinson, 1975), p. 77.
11. Ibid., p. 78.
12. B. L. Whorf, *Language, Thought, and Reality*, ed. J. B. Carroll (London, New York: Wiley, 1962), p. 55.
13. Ibid., pp. 57–64.
14. In addition to those born blind von Senden included those losing their sight in infancy, before concepts of space could develop.
15. M. von Senden, *Space and Sight*, trans. P. Heath (London: Methuen, 1960), p. 61.
16. Ibid., p. 63.
17. Ibid., p. 102.

18. Ibid., p. 106.
19. See also P. Winch, *The Idea of a Social Science and its Relation to Philosophy* (London: Routledge & Kegan Paul, 1961), p. 124.
20. W. Shakespeare, *The Merchant of Venice*, Act III, Sc.i.
21. It had been available ever since German measles had been recognised as a definite disease.
22. For example, Thomas Kuhn in *The Structure of Scientific Revolutions*, 2nd edn (University of Chicago Press, 1970).
23. See D. Stove, *Popper and After* (Oxford: Pergamon, 1982), Chapter 1.
24. R. Rorty, *Philosophy and the Mirror of Nature* (Oxford: Basil Blackwell, 1980), pp. 329–30.
25. Ibid., pp. 330–1.
26. Trigg, *Understanding Social Science*, p. 13.
27. L. Wittgenstein, *On Certainty* (Oxford: Basil Blackwell, 1974); Winch, *The Idea of a Social Science*; more explicitly D. Bloor, *Knowledge and Social Imagery* (London: Routledge & Kegan Paul, 1976).
28. Trigg, *Understanding Social Science*, p. 28.
29. Ibid., p. 28–9.
30. Ibid., p. 34.

3   FACTS AND THEORIES

1. W. Whewell (1774–1866), historian and philosopher of science; Master of Trinity College, Cambridge.
2. Whewell, quoted in *Theories of Scientific Method*, ed. E. H. Madden (London and Seattle: University of Washington Press, 1966), pp. 185–6.
3. Ibid., p. 188.
4. I. Scheffler, *Science and Subjectivity* (Indianapolis: Bobbs-Merrill, 1967), p. 13.
5. By contrast, if we saw a man apparently flying around the room we would be unlikely to accept what we saw as a plain matter of fact; we would suspect trickery or perhaps fear we were having some hallucination.
6. Quine, *Word and Object*, p. 3.
7. D. Thomas, *Naturalism and Social Science* (Oxford University Press, 1979), p. 34.
8. Ibid., p. 34.
9. The interaction between explanations offered by actors and observers is discussed further in Chapters 13 and 14.
10. Readers may think some of these facts, for example (7), need qualifying, but that is irrelevant here.

4   FACTS AND LANGUAGE

1. L. Wittgenstein, *Tractatus*, trans. D. F. Pears and B. F. McGuiness (London: Routledge & Kegan Paul, 1961), 5.62.
2. Ibid., 7.0.

3. Ibid., 2.0.
4. It is not the same in other respects; in particular an empirical fact is not a necessary truth but a contingent truth; it just 'happens to be the case' whereas a mathematical or a logical fact is necessarily true in relation to the conventions governing the use of the terms.
5. Events are not located in the same way as objects, for example the process of dying of lung cancer does not generally occur in one place. However, events can be *located* in a way that facts (particular and general) cannot be.
6. O'Connor, *Correspondence Theory of Truth*, p. 24.
7. W. Shakespeare, *Hamlet*, Act II, Sc.ii.
8. H. Putnam, *Meaning and the Moral Sciences* (London: Routledge & Kegan Paul, 1961), p. 99.
9. Ibid., p. 100.
10. It may be objected that animals and infants are aware of certain facts and so facts must exist for them, though not expressed in language. Thus the dog knows that his bone is buried under the tree, the baby knows that her mother will give her milk. They show that they possess simple concepts (see Chapter 2, p. 13) and they are aware of certain events and situations in the world. But facts are not themselves events or situations, neither are they awareness of events and situations, and it is doubtful whether the dog or the baby assess their knowledge as being a matter of fact *about the world*. It is in this sense, a sense that is the essence of facts, that facts exist only as expressed in language.
11. N. R. Hanson, *Perception and Discovery* (San Francisco: Freeman, Cooper, 1969), p. 185.
12. Ibid., p. 182.

5   TECHNICAL TERMS

1. The view that a genuine empirical explanation or theory must be subject to refutation by observation has been developed by Sir Karl Popper in *The Logic of Scientific Discovery* (London: Hutchinson, 1972) and in other writings of his. See also Chapter 9, note 1.
2. It is highly unlikely that any aspect of human behaviour will be satisfactorily explained by a single factor.
3. See J. Trusted, *The Logic of Scientific Inference* (London: Macmillan, 1979), Chapter 6, Section iv.
4. Other reasons were acknowledgement of the importance of observation and of planned experiments, less reliance on 'authority' and more communication between natural philosophers.
5. See also Hanson, *Perception and Discovery*, Chapter 3 and Trusted, *Logic of Scientific Inference*, Chapters 3, Section v.
6. There are claims that certain aspects of intelligence, for example the ability to make simple arithmetical calculations, can be measured objectively, and there are theories that a 'mental age' can be assigned to children on the basis of an analysis of the results of these tests. But there

remains much disagreement in this field and the concept of intelligence has not been clarified enough to enable it to become the subject of tests whose results are accepted by all those qualified to assess them.
7. G. Ryle, *Dilemmas* (Oxford University Press, 1979), pp. 90–1.

6   THE D-N MODEL AND THE CONCEPT OF LAW

1. See also Chapter 8.
2. In many cases there is not a full explanation given for the law is implied – assumed to be too well known to need stating (see p. 70).
3. Popper, *Logic of Scientific Discovery*, pp. 59–60.
4. C. G. Hempel, *Aspects of Scientific Explanation* (London: Collier-Macmillan, 1965), 'Studies in the Logic of Explanation', pp. 245 *et seq.*
5. A. Ryan, *The Philosophy of the Social Sciences* (London: Macmillan, 1984), pp. 52–3.
6. For an elementary discussion of this point see Trusted, *Logic of Scientific Inference*, Chapter 4, and Trusted, *Introduction to the Philosophy of Knowledge* (London: Macmillan, 1981), Chapter 7.
7. Winch, *The Idea of a Social Science*, p. 17.
8. Some philosophers, for example Leibniz, a near-contemporary of Descartes, have held that nothing occurs accidentally, that all is predetermined by the Deity. However, they can still distinguish laws of nature from 'accidental' true generalisations since, in respect of events conforming to laws, God has imposed a consistent pattern whereas the 'accidental' generalisations refer to facts not so ordered; God has there determined a lack of consistent pattern.
9. J. Kempthorn (1775–1838), *Hymns of Praise* (1796).
10. A. J. Ayer, *The Concept of a Person and Other Essays* (London: Macmillan, 1963), p. 220.
11. This 'necessity' is also referred to by Ayer; it is not logical necessity but factual necessity; hence the quotation marks.
12. Ayer, *Central Questions of Philosophy*, pp. 150–1.
13. Ryan, *Philosophy of the Social Sciences*, p. 200.
14. C. G. Hempel, 'The Logic of Scientific Explanation' in *Readings in the Philosophy of Science*, Feigl and Brodbeck (eds) (New York: Appleton-Century Crofts, 1953), p. 323.
15. N. Rescher, *Scientific Explanation* (New York: Free Press, 1970), p. 145.
16. See also Chapter 7, p. 65.
17. I. Scheffler, *The Anatomy of Inquiry* (London: Routledge & Kegan Paul, 1964), pp. 41–2.
18. For further discussion see Chapter 12.
19. Scheffler, *Anatomy of Inquiry*, pp. 54–5.
20. In explaining events relating to human behaviour (of individuals and in societies) we require a characteristic kind of conceptual understanding, *in addition to* that required for physical explanations of inanimate objects (and possibly plants and non-human animals). See also Chapters 1, 3 and 13.

7   ASPECTS OF EXPLANATION: HOW, WHY AND WHEREFORE

1. W. C. Salmon, 'Rational Prediction', *British Journal for the Philosophy of Science*, 32, No. 2 (1981), 115.
2. Chapter 2, p. 21.
3. Salmon, 'Rational Prediction', p. 125.
4. Von Wright, *Explanation and Understanding*, p. 6.
5. Ibid., p. 173.
6. See also Chapter 11.
7. Von Wright, *Explanation and Understanding*, p. 29.
8. Ibid., p. 30.

8   LAWS AND QUASI-LAWS

1. Trusted, *Logic of Scientific Inference*, Chapter 5, Section ii.
2. See Chapter 6, p. 57.
3. There are practical and theoretical difficulties associated with refuting statistical generalisations but these are irrelevant here.
4. See also Chapter 6, p. 57.
5. R. Carnap, *An Introduction to the Philosophy of Science*, M. Gardner (ed.) (New York: Basic Books, 1966), p. 7.
6. Rescher, *Scientific Explanation*, p. 179.
7. Thomas, *Naturalism and Social Science*, p. 18.

9   SPURIOUS EXPLANATIONS

1. See Karl Popper, *Conjectures and Refutations* (London: Routledge & Kegan Paul, 1961), pp. 33–7.
2. There are no university departments in Europe. For Paul Thagard this is the principal reason for rejecting astrology and for treating it as a pseudo-science. He produces interesting objections to the predictability and falsifiability criteria. See Paul R. Thagard, 'Why Astrology is a Pseudo-science', *Philosophy of Science Association*, I (1978).
3. There was a time when astrology would have been regarded as a genuine scientific subject. As Thagard says:

   One interesting consequence of the above criterion is that a theory can be scientific at one time but pseudo-scientific at another. In the time of Ptolemy or even Kepler astrology had few alternatives in the explanation of human personality and behaviour. (ibid., p. 229)

4. The current revival of interest in astrology supports Thagard's view of pseudo-science. He argues that it is up to individuals and to scientific communities to investigate alternatives and consider new interpretations of observations; he is not advocating a cultural relativism. His concluding paragraph is:

   In conclusion, I would like to say I think the question of what constitutes a pseudo-science is important. Unlike the logical positivists, I am not grinding an anti-metaphysical axe, and unlike Popper, I am not grinding an anti-Freudian or anti-Marxian one. My concern is social: society faces the twin problems of lack of public concern with the advancement of science, and lack of public concern with the important ethical issues now arising in science and technology, for example around the topic of genetic engineering. One reason for this dual lack of concern is the wide popularity of pseudo-science and the occult among the general public. Elucidation of how science differs from pseudo-science is the philosophical side of an attempt to overcome public neglect of genuine science. (Ibid., p. 230)

5. M. Shepherd, 'The Psycho-Historians', *Encounter* (March, 1979), p. 35.
6. Ibid., p. 36.
7. Ibid.
8. Ibid.
9. Ibid., p. 37.
10. Popper, *Conjectures and Refutations*, pp. 37–8.
11. H. J. Eysenck, 'How Scientific is Freudianism?', *Encounter* (January 1978), pp. 36–40.
12. S. Fisher and R. Greenberg, 'On Freud, Eysenck and Critical Passions', *Encounter* (April, 1978) pp. 91–92; H. J. Eysenck, letter in ibid.; R. Kee and H. J. Eysenck, correspondence, *Encounter* (May 1978), p. 92; P. Kline and H. J. Eysenck, correspondence, *Encounter* (July 1978), p. 93.
13. Eysenck, *Encounter* (January 1978), p. 38.
14. Ibid.
15. Trusted, *Logic of Scientific Inference*, Chapter 6, Section iv.

10 TELEOLOGICAL EXPLANATIONS

1. L. Wright, *Teleological Explanation* (London: University of California Press, 1976), p. 39. Note that L. Wright is to be distinguished from G. H. von Wright.
2. Ibid., p. 81.
3. See also Chapter 1, p. 7.
4. C. Taylor, *The Explanation of Behaviour* (London: Routledge & Kegan Paul, 1964), pp. 28–9.
5. Taylor points out that the antecedent is not the result which follows the event but the state of affairs obtaining prior to it, ibid., p. 16.
6. Ryan, *Philosophy of the Social Sciences*, pp. 114–15.
7. J. Trusted, *Free Will and Responsibility* (Oxford University Press, 1984), Chapter 17.
8. Ryan, *Philosophy of the Social Sciences*, pp. 119–21.
9. Wright, *Teleological Explanation*, p. 52.
10. Ibid., p. 53.
11. Ibid., pp. 53–4.

12. Ibid., pp. 122–9.
13. J. K. Derden Jr, 'Reasons, Causes, and Empathetic Understanding', *Philosophy of Science Association*, I (1978) 177.
14. Ibid., p. 178.
15. Von Wright, *Explanation and Understanding*, p. 27.
16. See also note 6, Chapter 6.
17. D. Hume, *Enquiries Concerning the Human Understanding*, L. A. Selby-Brigge (ed.), 2nd edn (Oxford: Clarendon Press, 1970), Part I, section viii para 71, p. 92.
18. Ibid., para 72, p. 93.
19. Wright, *Teleological Explanation*, p. 102.
20. N. Block, 'Introduction: What is Functionalism?', *Readings in Philosophy of Psychology*, N. Block (ed.) vol. I (1980), p. 172.
21. R. Cummins, 'Functional Analysis', in ibid., p. 185.
22. Von Wright, *Explanation and Understanding*, p. 157.
23. Ibid., p. 158.
24. Ibid., p. 133.
25. Ibid., pp. 134–5.
26. D. Harvey, *Explanation in Geography* (London: Edward Arnold, 1973), p. 444.
27. Ibid., p. 452.
28. Ibid., pp. 457–8.
29. Ibid., p. 446.
30. Ibid., pp. 435–6.
31. Ryan, *Philosophy of the Social Sciences*, p. 192.
32. Trigg, *Understanding Social Science*, p. 47.
33. Ibid., p. 50.
34. Ibid., pp. 50–1.

11   THE D-N MODEL AND THE HUMAN SCIENCES

1. These objections are listed and countered by E. Zilsel in his paper 'Physics and the Problems of Historico-sociological Laws', in *Readings in the Philosophy of Science*, Feigl and Brodbeck (eds).
2. Here we are not concerned with the metaphysical question as to whether our choice is free in some ultimate sense, or whether it is ultimately physically determined by our genetic makeup and past history. We are simply referring to situations where the individual is thought free to choose in a way that inanimate objects are not free.
3. Zilsel, 'Physics and the Problems of Historico-sociological Laws', p. 717.
4. Flew, *A Rational Animal*, p. 117.
5. Ryan, *Philosophy of the Social Sciences*, pp. 207–8.
6. Zilsel, 'Physics and the Problems of Historico-sociological Laws', p. 721.
7. See also Chapter 13, p. 133.
8. Winch, *The Idea of a Social Science*, pp. 83–4.
9. Ryan, *Philosophy of the Social Sciences*, pp. 146–7.
10. Trigg, *Understanding Social Science*, p. 64.

11. See also Chapters 1 and 2.
12. E. Nagel, 'The Logic of Historical Analysis', in Feigl and Brodbeck (eds) p. 697.
13. Rescher, *Scientific Explanation*, p. 187.
14. W. Dray, *Laws and Explanation in History* (Oxford: Clarendon Press), pp. 32–5.
15. Ryan, *Philosophy of the Social Sciences*, pp. 212–13.
16. Von Wright, *Explanation and Understanding*, p. 137.

12 INADEQUACIES OF THE D-N MODEL

1. D. M. Hausman, 'Causal Explanatory Asymmetry', *Philosophy of Science Association*, I (1982) 49.
2. See Putnam, *Meaning and the Moral Sciences*, p. 22 and P. Kitcher, 'Genes', *British Journal for the Philosophy of Sciences*, 33, no. 4 (1982) 337–59.
3. Quine, *Word and Object*, p. 16.
4. Ibid., p. 57.
5. Putnam, *Meaning and the Moral Sciences*, p. 37.
6. Ibid., p. 42.

13 THE OPERATION OF VERSTEHEN

1. T. Abel, 'The Operation called *Verstehen*', *Readings in the Philosophy of Science*, Feigl and Brodbeck (eds), p. 685.
2. Ibid., p. 687.
3. A. Schutz, *On Phenomenology and Social Relations* (University of Chicago Press, 1970), pp. 267–71.
4. Ibid., pp. 270–1.
5. Derden, 'Reasons, Causes and Empathetic Understanding', pp. 182–3.
6. Trigg, *Understanding Social Science*, p. 46; see also Chapter 1.
7. See also the brief reference at the end of Chapter 4 to Wittgenstein's view that language is significant in virtue of the rules that guide its use. There is also a discussion, supporting Wittgenstein's view that there cannot be a private language, in Winch's *The Idea of a Social Science*, Chapter I, sections 8 and 9.
8. L. Wittgenstein, *Philosophical Investigations*, trans. G. E. M. Anscombe (Oxford: Basil Blackwell, 1974), p. 258.
9. Ayer, 'Can There be a Private Language?', in *Concept of a Person*, p. 42.
10. Ibid., p. 43.
11. T. Nagel, 'What is it like to be a Bat?', *Readings in Philosophy of Psychology*, vol. I, Block (ed.), p. 163.
12. Ibid., p. 166.
13. O. Sacks, *Awakenings* (London: Picador, 1982), pp. 35–6.
14. David Hume asserted that human beings were nothing but bundles of perceptions and that there was no underlying substratum or 'self' holding

the bundle together. See D. Hume, *A Treatise of Human Nature*, L. A. Selby-Bigge (ed.) (Oxford: Clarendon Press, 1973), p. 252.
15. O. Sacks, *The Man who Mistook his Wife for a Hat* (London: Duckworth, 1985), p. 37.
16. Trigg, *Understanding Social Science*, pp. 41–2.
17. Ibid., p. 44.
18. Putnam, *Meaning and the Moral Sciences*, pp. 71–2.
19. The reference here is to Ernst Nagel, not the Thomas Nagel quoted (see notes 9 and 10).
20. Putnam, *Meaning and Moral Sciences*, p. 75.
21. Ibid., p. 77.
22. Of course, detailed explanations of the reasons for the faith of different individuals will vary. The operation of *Verstehen* is required for explanations of the behaviour of single individuals and groups of individuals (Societies).

14   VALUES AND EXPLANATION

1. There are several different forms of Marxism; we need not be concerned with that here.
2. M. Weber, ' "Objectivity" in Social Science and Social Policy', trans. E. A. Shils and H. A. French in G. Riley (ed.), *Values, Objectivity, and the Social Sciences* (London: Addison-Wesley, 1974), p. 75.
3. Ibid., p. 79.
4. A. Kaplan, 'Values in Inquiry', in Riley, *Values, Objectivity . . .*, p. 89.
5. Ibid., p. 98.
6. Ryan, *Philosophy of the Social Sciences*, p. 226.
7. M. Midgley, 'On Facts and Value', *Encounter* (October 1981), pp. 50–1.
8. Thomas, *Naturalism and the Social Sciences*, p. 314.
9. Ibid., pp. 134–5.
10. Normative standards are standards imposing rules and/or values.
11. S. G. Harding, 'Four Contributions Values can make to the Objectivity of Social Science', *Philosophy of Science Association* (1978), 1, 202.
12. Ibid.
13. B. Disraeli, *Sybil* (London: Folio Society, 1983), p. 83.
14. See also Chapter 3, p. 33.
15. Harding, 'Four Contributions', p. 203.
16. Ibid., p. 205.
17. R. Nisbet, 'Subjective Si! Objective No!', in Riley, *Values, Objectivity . . .*, pp. 16–17.
18. Harding, 'Four Contributions', p. 207.
19. See Chapter 13, p. 139, quote 21.
20. Animals may be corrected and even punished in training, but this is a matter of conditioning; we do not regard animals as morally responsible creatures.

Bibliography

ALEXANDER, P., *Sensationalism and Scientific Explanation* (London: Routledge and Kegan Paul, 1963)

AYER, A. J., *The Central Questions of Philosophy* (London: Weidenfeld and Nicolson, 1973)

——, *The Concept of a Person and Other Essays* (London: Macmillan, 1963)

BLOCK, N. (ed.), *Readings in the Philosophy of Psychology*, vol. I (London: Methuen, 1980); vol. II (London: Methuen, 1981)

BLOOR, D., *Knowledge and Social Imagery* (London: Routledge and Kegan Paul, 1976)

British Journal for the Philosophy of Science, 32 (1981); 33 (1982)

CARNAP, R., *An Introduction to the Philosophy of Science*, M. Gardner (ed.) (New York: Basic Books, 1966)

DISRAELI, B., *Sybil* (London: Folio Society, 1983)

DRAY, W., *Laws and Explanations in History* (Oxford: Clarendon Press, 1957)

Encounter (January, April, May, July 1978); (March 1979); (October 1981)

FEIGL, H. and BRODECK, M. (eds), *Readings in the Philosophy of Science* (New York: Apple-Century Crofts, 1953)

FLEW, A., *A Rational Animal* (Oxford: Clarendon Press, 1978)

GOODMAN, N., *Fact, Fiction and Forecast* (New York: Bobbs-Merrill, 1973)

GREGORY, R., *Eye and Brain* (London: Weidenfeld and Nicolson, 1966)

HANSON, N. R., *Perception and Discovery* (San Francisco: Freeman, Cooper, 1969)

HARVEY, D., *Explanation in Geography* (London: Edward Arnold, 1973)

HEMPEL, C. G., *Aspects of Scientific Explanation* (London: Collier-Macmillan, 1965)

HUME, D., *Enquiries Concerning the Human Understanding*, L. A. Selby-Bigge (ed.), 2nd edn (Oxford: Clarendon Press, 1970)

——, *A Treatise of Human Nature*, L. A. Selby-Bigge (ed.) (Oxford: Clarendon Press, 1973)

KEENE, G. B., *Language and Reasoning* (Toronto, London and New York: D. van Nostrand, 1961)

KUHN, T. S., *The Structure of Scientific Revolutions*, 2nd edn (University of Chicago Press, 1970)

MADDEN, E. H. (ed.), *Theories of Scientific Method* (London and Seattle: University of Washington Press, 1966)

MERTON, R. K., *Social Theory and Social Structure* (London: Collier-Macmillan, 1975)

O'CONNOR, D. J., *The Correspondence Theory of Truth* (London: Hutchinson, 1975)

POPPER, K., *Conjectures and Refutations* (London: Routledge & Kegan Paul, 1961)

———, *The Logic of Scientific Discovery* (London: Hutchinson, 1972)
Proceedings of the Philosophy of Science Association:

———, (1978), I, P. D. Asquith and I. Hacking (eds)

———, (1980), I, P. D. Asquith and R. M. Giere (eds)

———, (1982), I, P. D. Asquith and T. Nickles (eds)
(East Lansing, Michigan: Philosophy of Science Association)

PUTNAM, H., *Meaning and the Moral Sciences* (London: Routledge & Kegan Paul, 1961)

QUINE, W. V. O., *Word and Object* (Cambridge, Mass.: MIT Press, 1964)

RESCHER, N., *Scientific Explanation* (New York: Free Press, 1970)

RILEY, G., (ed.), *Values, Objectivity and the Social Sciences* (London: Addison-Wesley, 1974)

RORTY, R., *Philosophy and the Mirror of Nature* (Oxford: Basil Blackwell, 1980)

RYAN, A., *The Philosophy of the Social Sciences* (London: Macmillan, 1984)

RYLE, G., *Dilemmas* (Oxford: Oxford University Press, 1979)

SACKS, O., *Awakenings* (London: Picador, 1982)

———, *The Man who Mistook His Wife for a Hat* (London: Duckworth, 1985)

SCHEFFLER, I., *The Anatomy of Inquiry* (London: Routledge & Kegan Paul, 1964)

———, *Science and Subjectivity* (Indianapolis: Bobbs-Merrill, 1967)

SCHUTZ, A., *On Phenomenology and Social Relations* (University of Chicago Press, 1970)

VON SENDEN, M., *Space and Sight*, trans. P. Heath (London: Methuen, 1960)

STOVE, D., *Popper and After* (Oxford: Pergamon, 1982)

TAYLOR, C., *The Explanation of Behaviour* (London: Routledge & Kegan Paul, 1964)

THOMAS, D., *Naturalism and Social Science* (Oxford University Press, 1979)

TRIGG, R., *Understanding Social Science* (Oxford: Basil Blackwell, 1985)

TRUSTED, J., *Free Will and Responsibility* (Oxford University Press, 1984)

———, *An Introduction to the Philosophy of Knowledge* (London: Macmillan, 1981)

———, *The Logic of Scientific Inference* (London: Macmillan, 1979)

WHORF, B. L., *Language, Thought and Reality*, ed. J. B. Carroll (London, New York: Wiley, 1962)

WINCH, P., *The Idea of a Social Science and its Relation to Philosophy* (London: Routledge & Kegan Paul, 1961)

WITTGENSTEIN, L., *On Certainty* (Oxford: Basil Blackwell, 1974)

———, *Philosophical Investigations*, trans. G. E. M. Anscombe (Oxford: Basil Blackwell, 1974)

———, *Tractatus*, trans. D. F. Pears and B. F. McGuiness (London: Routledge & Kegan Paul, 1961)

WRIGHT, L., *Teleological Explanation* (London: University of California Press, 1976)

VON WRIGHT, G. H., *Explanation and Understanding* (London: Routledge & Kegan Paul, 1971)

Index